DIGITAL EQUIPMENT
TROUBLESHOOTING

DIGITAL EQUIPMENT TROUBLESHOOTING

M. NAMGOSTAR

RESTON PUBLISHING COMPANY, INC.
A Prentice-Hall Company
Reston, Virginia

Technical Drawings by Robert Mosher

Library of Congress Cataloging in Publication Data

Namgostar, M
Digital equipment troubleshooting.

Includes index.
1. Electronic apparatus and appliances—Maintenance and repair. 2. Digital electronics. I. Title.
TK7868.D5H4   621.3815′028   77-23148
ISBN 0-87909-201-7

© 1977 by Reston Publishing Company, Inc.
A Prentice-Hall Company
Reston, Virginia 22090

All rights reserved. No part of this book may be reproduced in any way, or by any means, without permission in writing from the publisher

10 9 8 7 6 5 4 3 2 1

PRINTED IN THE UNITED STATES OF AMERICA

D
621.3815
NAM

CONTENTS

Preface ix

1 • Decision-Making Elements 1
    1.1 AND Gates 1
    1.2 OR Gates 2
    1.3 Inverter Characteristics 3
    1.4 Applications of Logic Gates 4
    1.5 Timing Diagrams 5
    1.6 Logic Circuits 7
    1.7 NAND and NOR Gates 10
    1.8 Dual Function of the NAND Gate 11
    1.9 Dual Function of the NOR Gate 11
    1.10 Interchange of Symbols 12
    1.11 EXCLUSIVE OR Gate 13
    1.12 Memory Elements 14
    1.13 RS Latch 14
    1.14 RS Latch from NAND Gates 18
    1.15 Race Condition 18
    1.16 Other Flip-Flops 19
    1.17 Clock, Preset, and Clear Signals 19

2 · Computer and Calculator Organization 22
- 2.1 General Considerations 22
- 2.2 Memory and ALU Operation 24

3 · Basic Digital Test Instruments 47
- 3.1 Troubleshooting With a Logic Probe 47
- 3.2 Troubleshooting With the Logic Pulser and Logic Probe 54
- 3.3 Troubleshooting With the Logic Clip 61
- 3.4 Troubleshooting With the Logic Comparator 63
- 3.5 Digital Troubleshooting Techniques 65
- 3.6 Digital Symptom Trouble Analysis 70
- 3.7 Multi-Family Logic Probes and Pulsers 76
- 3.8 Oscilloscopes 77

4 · Preliminary Troubleshooting Approach 82
- 4.1 Basic Troubleshooting Procedure 82
- 4.2 Procedural Principles 88

5 · Functional Analysis of a Digital System 95
- 5.1 System Evaluation 95
- 5.2 Electrical Analysis 104

6 · Systems Troubleshooting Procedures 114
- 6.1 Sequential Plan of Attack 114
- 6.2 Pattern Consistencies 116
- 6.3 Logic Troubleshooting 117
- 6.4 Eight Basic Rules for Digital Troubleshooting 120
- 6.5 Development of Troubleshooting Expertise 121
- 6.6 Job Descriptions 126
- 6.7 Symptom Analysis 127
- 6.8 Diagnostic Program 130
- 6.9 Intermittent Trouble 131
- 6.10 Troubleshooting Shortcuts 132
- 6.11 Learning from the Experience of Others 133
- 6.12 Additional Aids 134

7 · Digital Logic Analyzers 139
- 7.1 General Considerations 139
- 7.2 Principles of Functional Analysis 139
- 7.3 Logic Analyzers 140

7.4 Triggering Serial Data   142
7.5 Triggering Parallel Data   144
7.6 Logic State Analyzer Characteristics   144
7.7 Logic Analyzer Operation and Application   146
7.8 Pattern Analyzer Operation   149

8 • Testing of Digital and Printed-Circuit Boards   150

8.1 Functional Testing and Fault Isolation   150
8.2 Fault Isolation   154
8.3 Digital Test Systems   155
8.4 Probe Hardware   159

9 • Test Models of Printed-Circuit Cards   161

9.1 Primitive Elements   161
9.2 Macro Characterization   167
9.3 Tristate Modeling Techniques   167

10 • Digital Logic Simulation in Test Procedures   175

10.1 Simulation of a Model   175
10.2 Races, Critical Races, and Hazards   179
10.3 Model Verification   182
10.4 Board Modeling   185

11 • Modeling and Simulating Faults   188

11.1 General Considerations   188
11.2 Detection of Faults   193
11.3 Levels of Fault Insertion   196
11.4 Fault Machines   197
11.5 Logically Equivalent Faults   197
11.6 Fault Signatures   198
11.7 Stop on First Fail   200
11.8 Multiple Fault Sites   201
11.9 Trade-offs on Selecting Faults   202
11.10 Possible Fault Detection   204
11.11 Testing Memory Elements   205

12 • Test Pattern Generation   207

12.1 Path Sensitization   207
12.2 Initialization   210
12.3 Fault Location and Simulation Results   214

12.4 Post Processors 215
12.5 Interface Adapters 215
12.6 Test System Operating Software 216
12.7 Fault Location 217
12.8 Feedback Loops 219
12.9 A Fault Isolation Probing Algorithm 220
12.10 Parallel Drivers 223
12.11 Internal Device Feedback 224
12.12 Pulse Response 225
12.13 Operator-Directed Probing 226
12.14 Validation 227

Appendices
I • **Troubleshooting the Intel 4004 Microprocessor Systems 229**
II • **Troubleshooting the Motorola M6800 Microprocessor System 241**
III • **Troubleshooting the Fairchild F8 Microprocessor Systems 250**

Glossary 259

Index 273

PREFACE

With the rapid advance of digital technology in recent years, coupled with the development of specialized digital test equipment, a need has arisen for a relevant state-of-the-art troubleshooting text and guidebook. This book covers a wide range of digital equipment from clocks to computers. Preliminary analysis of trouble symptoms is followed by explanation of various techniques that may be utilized to narrow down the possible faults. Use of logic probes, logic pulsers, logic clips, logic comparators, oscilloscopes, and data analyzers is discussed and illustrated. The text starts with troubleshooting procedures for simple digital equipment and progresses functionally to comparatively complex equipment. This is essentially a "how to" book that addresses itself primarily to the professional digital technician and to the vocational student.

Digital Equipment Troubleshooting minimizes theoretical considerations; the author assumes that the reader has completed basic training in digital theory. However, the reader need not have any on-the-job troubleshooting experience. This text is designed to bridge the gap between classroom theory and field or shop hands-on activities. Thus, this book will find equal utility in the vocational school and in industry. It is one of the few books written for the practicing digital technician. The author wishes to acknowledge his indebtedness to the engineers and executive personnel at IBM, Hewlett-Packard, and Digital Equipment Corporation, who were most cooperative in providing technical advice and material. He also wishes to thank Tektronix, Inc., for their kind permission to reproduce excerpts from their company publications.

This teaching tool is the outcome of extensive teaching experience, both on the part of the author and of his fellow instructors at San Jose City College, who have offered valuable advice and constructive criticisms. In a significant sense, this is a team-effort accomplishment, and it is appropriate that this work be dedicated as a teaching tool to the instructors and students of our community colleges, vocational schools, and technical institutes.

M. N.

1 •
DECISION-MAKING ELEMENTS

1.1 AND Gates

The AND gate is a device whose output is a logic 1 if both of its inputs are logic 1. If only one input is a 1, with the other input a logic 0, the output will be a 0. This gate is shown by the symbol in (a) of Figure 1-1, where the two inputs are on the left, marked A and B, and the output is on the right, marked C. To visualize the AND gate, it is helpful to use an analogy—the light-bulb circuit in (b) of Figure 1-1. In this circuit, both switches A and B must be closed for the light bulb to be on. If only one of the switches is closed, the light bulb will be off. The two switches are, therefore, analogous to the AND gate inputs A and B, while the bulb corresponds to the output.

Various combinations of input states of an AND gate and its response to these inputs can be expressed by making up a table, as in (c) of Figure 1-1. The two columns on the left show the states of the inputs to the AND gate, and the column on the right shows the corresponding output. If you relate this table to the lamp circuit, a 0 represents the lamp off or open-switch condition, and a 1 represents the lamp on or closed-switch condition. If you read horizontally along the lines of this table, you will see the response of the output (or lamp) to all combinations of the inputs. This table is called a *truth table*.

Another method of representing the operation of an AND gate is

2 · Decision-Making Elements

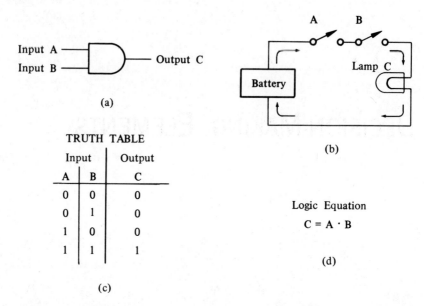

FIGURE 1-1 AND gate characteristics. (a) Logic symbol; (b) operating principle; (c) logic truth table; and (d) logic equation.

called the *Boolean logic equation,* or simply the *logic equation.* The symbol used to represent the AND function is [·]. The AND function referred to in the previous example can be written $A \cdot B$ (A and B) as in (d) of Figure 1-1, or simply as AB. The full logic equation that relates the output C to the inputs A and B is $C = A \cdot B$. It reads *C equals A and B*. This equation means that both A and B must be logic 1 if C is to be logic 1.

1.2 OR Gates

The OR gate is a device whose output is a logic 1 if either or both of its inputs are a logic 1. The OR gate is shown by the symbol in (a) of Figure 1-2, with the two inputs A and B again on the left and the output C on the right. To visualize the OR gate, a light-bulb circuit can be used, with the switches connected in parallel. In turn, the bulb can be turned on by closing either of the switches A or B, or both. The truth table for the OR gate is shown in (c) of Figure 1-2, and the logic equation is $C = A + B$. This equation reads: *C equals A or B*. Note the differences between the truth tables and the logic equations of the OR and AND gates. The symbol for the OR function is (+), but this

1.3 Inverter Characteristics · 3

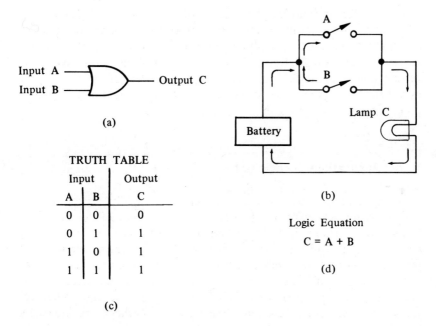

FIGURE 1-2 OR gate characteristics. (a) Logic symbol; (b) operating principle; (c) logic truth table; and (d) logic equation.

should not be confused with the plus sign of mathematics; the Boolean symbols have functions that are unique to Boolean algebra.

1.3 Inverter Characteristics

The third and most simple element of digital logic is the inverter, shown in Figure 1-3. (It is also known as the NOT function.) The in-

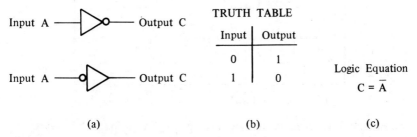

FIGURE 1-3 Inverter characteristic. (a) Logic symbols; (b) truth table; and (c) logic equation.

4 · Decision-Making Elements

verter is different from the AND and OR gates, in that it has only a single input. As a result, it does not perform a decision-making function that is dependent on a combination of inputs. Instead, the inverter simply converts a logic 1 at its input to a logic 0 at its output and, conversely, converts a logic 0 to a logic 1. The inverter can be represented by either of the symbols shown in (a) of Figure 1-3, and its logic equation $C = \overline{A}$ is read: *C equals not A (or A not)*.

1.4 Applications of Logic Gates

AND, OR, and inverter circuits are most commonly found in integrated circuits, as depicted in Figure 1-4. Sealed inside of the integrated circuit (IC) package is a chip of silicon that has been processed in a manner that provides a microscopic pattern of resistors, diodes, and transistors on it. These miniature components make up the single gates in the package, which can be repeated many times on the same chip to make more complicated logic functions, to make several individual gates. The connecting pins extending from the package are internally connected to the inputs and outputs of the gates and are used for making external connections to the gates. To identify each pin for the purpose of external connections, the pins are numbered and a small diagram (called a *pin diagram*) is provided by the manufacturer with the integrated circuit. Integrated circuits are usually mounted on plug-in circuit boards called *modules*. Figure 1-5 shows a small digital computer, and Figure 1-6 shows the modules used in the computer.

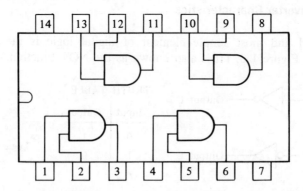

FIGURE 1-4   Arrangement of an IC package with four AND gates.

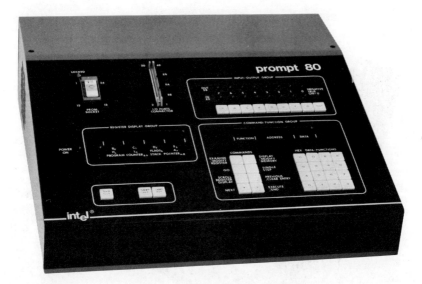

FIGURE 1-5 A small kit-type computer. (*Courtesy, intel*)

1.5 Timing Diagrams

When a logic gate is performing a useful function in a circuit, its inputs can change, and its output will react to these changes according to the truth tables for that gate. It is often quite useful to have a symbolic representation for the logic states, because they change with time. A convenient method for doing this is to draw a timing diagram. In Figure 1-7, the operation of an AND gate is shown in a timing diagram for an arbitrary sequence of high and low logic signals (voltages) applied to the A and B inputs. Input A, input B, and output C of the AND gate are each represented by a continuous line that proceeds from left to right as time progresses. Each of these lines has two possible levels, corresponding to whether the particular input or output is logic 1 (high) or logic 0 (low) at a particular time. Note that output C corresponds to the level predicted by the truth table and logic equation of Figure 1-1.

The main purpose of a timing diagram is to show what the conditions in a logic circuit are at any one particular time. By using timing lines, it is possible to oversee all inputs and outputs simultaneously. If any input or output line is displayed on an oscilloscope screen, you

6 · Decision-Making Elements

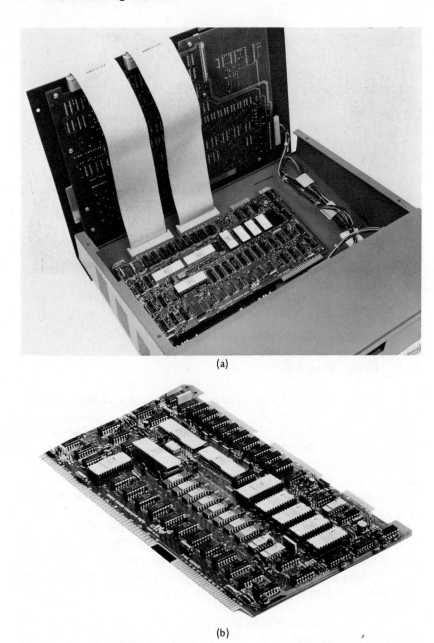

FIGURE 1-6 Modules used in the computer illustrated in Figure 1-5: (a) cover open; (b) computer board; (c) close-up of memory chip. (*Intel Corporation*)

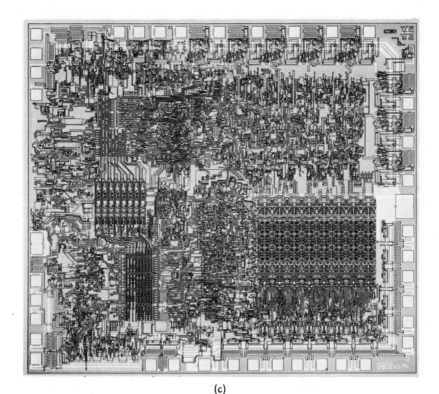

(c)

would see a display very much like the timing diagram shown in Figure 1-7. Logic analyzers that display a number of input and output lines simultaneously on an oscilloscope screen are used to troubleshoot malfunctioning digital circuitry.

1.6 Logic Circuits

Two AND gates can be interconnected as shown in Figure 1-8. To turn on the first gate, we need high input at both A and B, but to turn on the second gate, we need high inputs at D, A, and B—a total of three high inputs. These two gates can be considered as one logic circuit, and the logic equation and truth table can be written for the circuit as a whole, as in (b) of Figure 1-8. The truth table is constructed in the same manner as has been explained, except that there are now three

8 · Decision-Making Elements

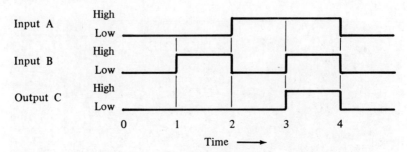

FIGURE 1-7 A timing diagram for an AND gate.

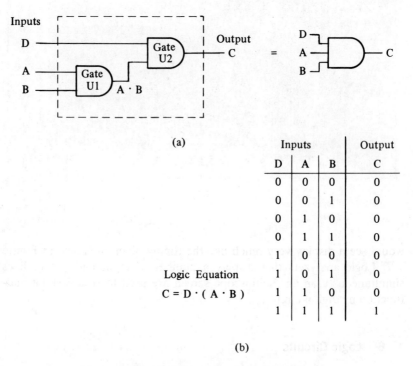

(a)

Logic Equation
$C = D \cdot (A \cdot B)$

Inputs			Output
D	A	B	C
0	0	0	0
0	0	1	0
0	1	0	0
0	1	1	0
1	0	0	0
1	0	1	0
1	1	0	0
1	1	1	1

(b)

FIGURE 1-8 A three-input AND gate circuit.

inputs to deal with instead of just two, and there are eight possible combinations of inputs instead of just four.

Consider the logic circuit depicted in Figure 1-9. A high logic level at D and either A or B will yield a logic-high output C. The truth

1.6 Logic Circuits · 9

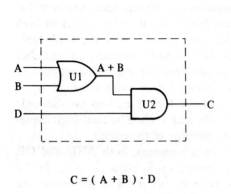

Inputs			Output
A	B	D	C
0	0	0	0
0	0	1	0
0	1	0	0
0	1	1	1
1	0	0	0
1	0	1	1
1	1	0	0
1	1	1	1

$C = (A + B) \cdot D$

FIGURE 1-9 A three-input AND/OR gate circuit.

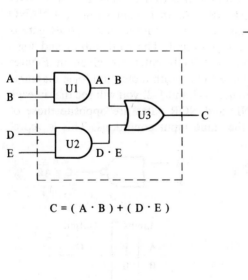

Inputs				Output
A	B	D	E	C
0	0	0	0	0
0	0	0	1	0
0	0	1	0	0
0	0	1	1	1
0	1	0	0	0
0	1	0	1	0
0	1	1	0	0
0	1	1	1	1
1	0	0	0	0
1	0	0	1	0
1	0	1	0	0
1	0	1	1	1
1	1	0	0	1
1	1	0	1	1
1	1	1	0	1
1	1	1	1	1

$C = (A \cdot B) + (D \cdot E)$

FIGURE 1-10 A four-input AND/OR gate circuit.

10 · Decision-Making Elements

table and logic equation are a restatement of this fact. Observe the circuit shown in Figure 1-10 in which high inputs at A and B or high inputs at D and E will yield a high output at C. Note the truth table and logic equation for this circuit. In this same manner, any number of gates can be combined and then analyzed for the characteristics of the resulting circuit. This analysis is simplified by using the logic equations to record intermediate values of the logic functions on the logic diagram, as shown in Figures 1-8, 1-9, and 1-10. The final logic equation, then, completely describes the operation of the circuit.

Although two-input gates are quite common, both AND and OR gates can have multiple inputs, numbering up to eight or more. In the case of the AND gate, the output is high only if all of the inputs are high. For the OR gate, the output is high whenever any one or more of the inputs are high. As an example, the two gates in (a) of Figure 1-8 can be and usually are replaced by a single three-input gate.

1.7 NAND and NOR Gates

Two types of logic gates often found in digital circuits are the NAND (NOT AND) and NOR (NOT OR). As their names imply, a NAND gate is an AND gate with an inverter on its output, and a NOR gate is an OR gate with an inverter on its output. The truth tables and logic equations for two-input NAND and NOR gates are given in Figures 1-11 and 1-12. In comparing these to the truth tables and logic diagrams of the AND and OR gates in Figures 1-1 and 1-2, you will see that in each case, the outputs of the NAND and NOR gates are opposite those of the AND and OR gates for the same input conditions. This is repre-

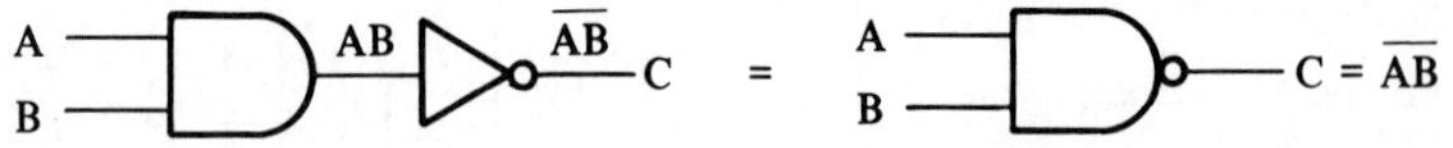

$C = \overline{A \cdot B}$

(The logic equation for the NAND gate reads: C equals A and B NOT)

Inputs		Output
A	B	C
0	0	1
0	1	1
1	0	1
1	1	0

FIGURE 1-11 Characteristics of a NAND gate.

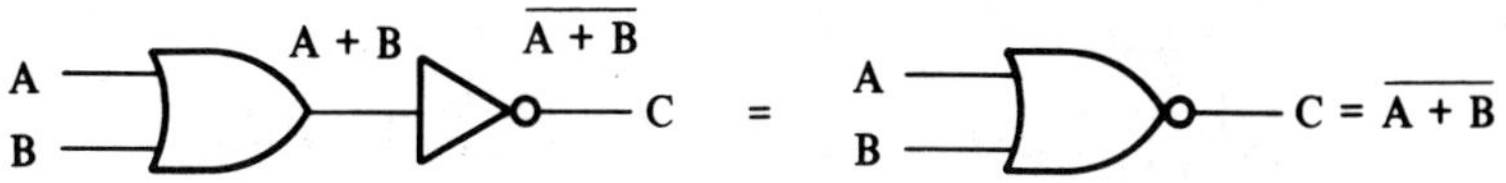

1.9 Dual Function of the NOR Gate · 11

$C = \overline{A + B}$

(The logic equation reads:
C equals A or B NOT)

Inputs		Output
A	B	C
0	0	1
0	1	0
1	0	0
1	1	0

FIGURE 1-12 Characteristics of a NOR gate.

sented in the logic equation by placing a NOT symbol (a bar) over $A \cdot B$ and $A + B$; hence, $\overline{A \cdot B}$ and $\overline{A + B}$. As shown in Figures 1-11 and 1-12, the inversion in the NAND and NOR gate is represented by a small circle at the output of the device. The NAND and NOR functions are widely used, because the inversion function is often required when building a digital circuit, and because both the NAND and NOR gates can be used in two ways, as will be explained in Secs. 1-8 and 1-9.

1.8 Dual Function of the NAND Gate

If you look at the truth table for the NAND gate shown in Figure 1-13, it is apparent that by inverting (changing 1's to 0's, and 0's to 1's) all the inputs to the NAND gate, we have exactly the truth table of the OR gate. In other words, if the inputs (A, B) applied to the NAND gate are already inverted ($\overline{A}$, $\overline{B}$), it will perform the OR function. If these inputs are not inverted, it will perform as an AND gate with the output inverted (see Figure 1-13). This dual function is represented by having two symbols for the NAND gate.

1.9 Dual Function of the NOR Gate

Inspection of the NOR truth table in Figure 1-14 shows that inverting all the inputs in the NOR truth table yields the AND truth table. A NOR gate can, therefore, not only be used as an OR gate with an inverter on the output, but also as an AND gate if the inputs (A, B)

12 · Decision-Making Elements

NAND				OR		
Inputs		Output		Invert inputs		Output
A	B	C		$\bar{A}$	$\bar{B}$	C
1	1	0		0	0	0
1	0	1		0	1	1
0	1	1		1	0	1
0	0	1		1	1	1

Invert inputs

Nand symbols

THEREFORE

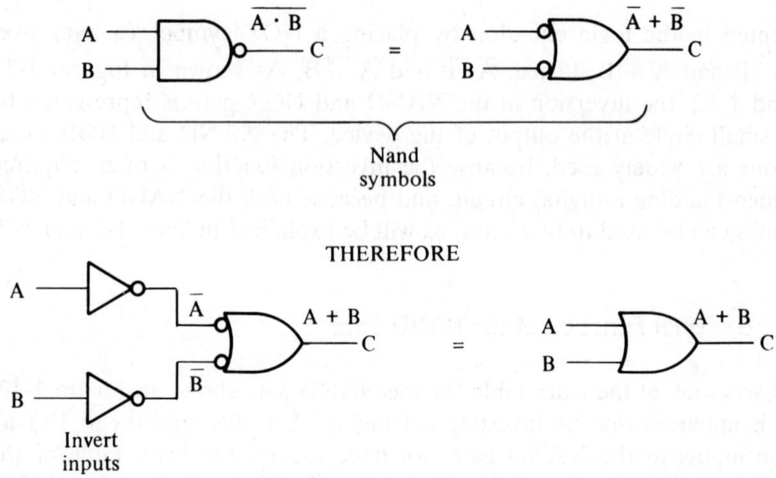

FIGURE 1-13 Two forms of NAND symbol.

are already in the inverted form ($\bar{A}$, $\bar{B}$). The two symbols for a NOR gate are shown in Figure 1-14; the small circle indicates the inversion function.

1.10 Interchange of Symbols

It has been seen that both the NAND gate and the NOR gate can be drawn two ways. The alternate ways of drawing the same gate symbol emphasize either the AND-type or the OR-type function. The type of symbol employed is dependent on how the gate is used in the circuit

1.11 EXCLUSIVE OR Gate · 13

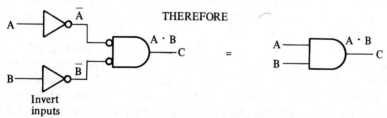

NOR				AND		
Inputs		Output		Inverted inputs		Output
A	B	C		A	B	C
1	1	0		0	0	0
1	0	0		0	1	0
0	1	0		1	0	0
0	0	1		1	1	1

FIGURE 1-14 Two forms of NOR symbol.

and is intended to make the logic diagram easier to understand. The ability of the NAND and the NOR gates to be used in more than one way very often makes a circuit function possible with a significantly smaller number of IC packages, compared to a solution that uses only AND/OR gates and inverters.

1.11 EXCLUSIVE OR Gate

The EXCLUSIVE OR gate is shown in Figure 1-15. It has a logic high output when either of its inputs is high, but not when both are high.

14 · Decision-Making Elements

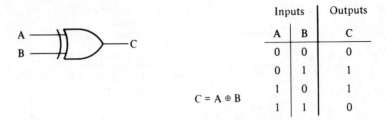

FIGURE 1-15 EXCLUSIVE OR gate and truth table.

Thus, it functions like an OR gate, excluding the last state of an OR truth table (Figure 1-2); in this state the output is inverted. Note that the logic equation introduces a new symbol $\oplus$ called EXCLUSIVE OR. This gate is quite useful, since its output is high only when its inputs are different. Figure 1-16 presents a reference chart of decision-making elements.

1.12 Memory Elements

AND and OR gates are commonly interconnected to provide memory: the arrangement can remember a signal of logic 1 or logic 0 level that has been connected to its input and can make the fact available at its output. The output of a memory circuit is therefore determined by the past inputs as well as the present inputs, whereas AND and OR gates simply make decisions based on their present inputs. An elementary memory device can be constructed from a single OR gate, as shown in Figure 1-17. In analyzing the operation of the element in Figure 1-17, assume that the output Q and input A are initially at logic 0 and, therefore, that the B input (which feeds back from output Q) is also at logic 0. If the signal at A is changed to a logic 1, the output will go to logic 1 also. However, if input A now returns to logic 0, the output will not change, because input B will still be at logic 1 and will keep the output at logic 1. Thus, the OR gate remembers that it received a logic 1 level, and the only way to erase that fact from this memory is to physically disconnect the wire between Q and B and place B at logic 0.

1.13 RS Latch

If the memory in Figure 1-17 has to remember only one event (one logic 1) in its lifetime and does need to be erased and reused for another event, it is adequate. In real situations, however, that is not the

1.13 RS Latch · 15

LOGIC SYMBOLS		TRUTH TABLE		
		Input A	Input B	Output C
OR				
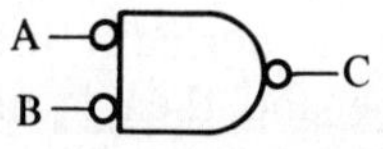 A—o⊐&—C B—o	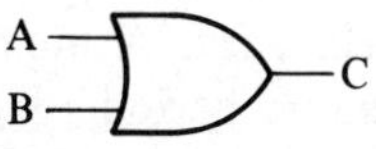 A—⊃≥—C B—	0	0	0
		0	1	1
		1	0	1
		1	1	1
AND				
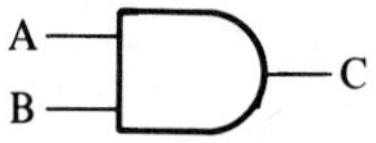 A—⊃&—C B—	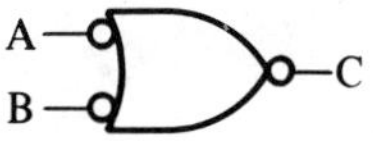 A—o⊃≥o—C B—o	0	0	0
		0	1	0
		1	0	0
		1	1	1
NAND				
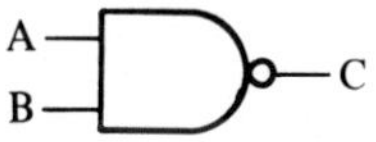 A—⊃&o—C B—	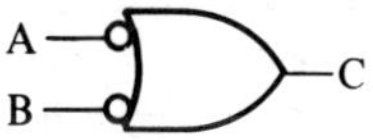 A—o⊃≥—C B—o	0	0	1
		0	1	1
		1	0	1
		1	1	0
NOR				
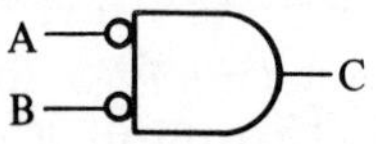 A—o⊃&—C B—o	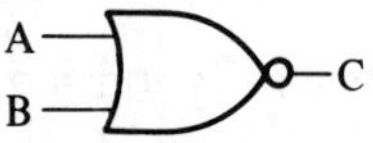 A—⊃≥o—C B—	0	0	1
		0	1	0
		1	0	0
		1	1	0
EXCLUSIVE OR				
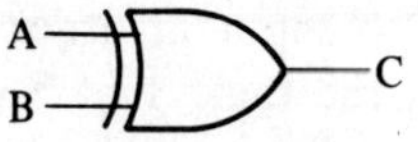 A—⊃=1⊃—C B—		0	0	0
		0	1	1
		1	0	1
		1	1	0

Note: On all TTL gates any unconnected input is interpreted as a logic 1. (An unconnected input is referred to as a floating input.)

FIGURE 1-16 Reference chart of decision-making elements.

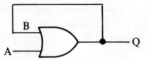

FIGURE 1-17   Example of a basic memory element.

case. Instead, a memory that can be erased or reset when it no longer needs to remember an event is necessary. A memory is normally used to store information (data) for some period of time and then is discarded. To implement this type of memory, two NOR gates are utilized, as shown in Figure 1-18. It is called an RS (reset-set) flip-flop or RS latch and is the most basic form of the class of circuits called *flip-flops*. At this point, we may consider that *flip-flop* and *latch* are equivalent terms.

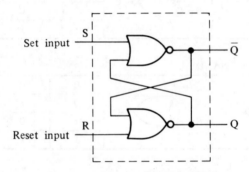

FIGURE 1-18   Basic RS latch.

To analyze the RS latch, begin by raising the *set input S* to a logic 1 level, while holding the *reset input R* to a logic 0. First of all, $\bar{Q}$ output goes to the logic 0 state. But logic 0 at Q is connected to the lower gate, so the lower gate now has two logic 0 inputs—and its output Q goes to logic 1. The logic 1 at Q is connected back to the input of the upper gate, and the upper gate now has two logic 1 inputs. Being a NOR gate, it needs only one high input in order to maintain its output in the logic 0 state. Therefore, the logic 1 at S can now become a logic 0, and the $\bar{Q}$ output will not change. If the logic 1 at S goes to a logic 0, all conditions remain the same. The S input is used to activate (or set) the circuit. As a result, the gates have latched themselves in certain states, and, as long as the R input remains a logic 0, nothing will change. Accordingly, this device is called a *latch*.

1.13 RS Latch

To reset the circuit, raise R to a logic 1. Q will become a logic 0 and $\bar{Q}$ a logic 1. Thus, the circuit has been reset, and its memory has been erased. The RS latch has the property that whenever the set input is raised to a logic 1, the Q output will become latched in a Q = 1 state, and whenever the reset input is raised to a logic 1, Q will become a logic 0. If you alternately raise one input and then the other, Q and $\bar{Q}$ will each alternate between logic 1 and logic 0 states. There is a final condition that should be considered: both inputs R and S high simultaneously. Since any one high input to the NOR gates will cause the output to be logic 0, both outputs will be logic 0 under these conditions. This is a special state of the RS latch that should be avoided. However, it does not change the basic function of the circuit, which is that of remembering a logic 1 or 0 at its inputs.

For simplicity, the RS latch is shown as a rectangle with labeled inputs and outputs, as in Figure 1-19. This representation is standard for all flip-flops and shows the flip-flop as a simple memory element whose essential purpose is to store a logic 1 or 0. This stored logic level is always available at the Q output of the flip-flop, and its *complement* (opposite) is present on $\bar{Q}$ output. That is, if Q is a logic 1, $\bar{Q}$ is a logic 0, and vice versa. If $\bar{Q}$ is a logic 0, Q is a logic 1. The stored logic level can be removed by storing another logic level in its place. If a logic 1 is stored in the flip-flop (Q = 1), the flip-flop is referred to as being set. If a logic 0 is stored (Q = 0), the flip-flop is referred to as being reset.

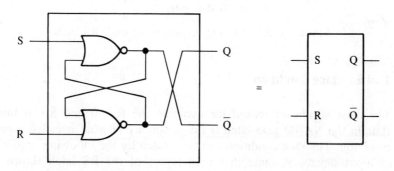

(a) Functional representation of the R − S latch as a simple memory element

(b) Standard Flip-Flop symbol as it is used in logic diagrams

FIGURE 1-19 Basic flip-flop symbol.

1.14 RS Latch from NAND Gates

A basic RS latch circuit can also be constructed from NAND gates, as shown in Figure 1-20. The analysis of this circuit is very similar to that of the NOR gate circuit. The two types of RS latches are the same in their ability to store a logic level. The NAND gate implementation of the latch is set and reset using logic 0's instead of logic 1's, as indicated by the inversion bars on the $\bar{R}$ and $\bar{S}$ inputs in Figure 1-20. To store a logic 1 at Q (logic 0 at $\bar{Q}$), the condition $\bar{S} = 0$ and $\bar{R} = 1$ must be at the inputs. This condition is remembered when the inputs return to $\bar{R} = 1$ and $\bar{S} = 1$. Conversely, a logic 0 is stored at Q (logic 1 at $\bar{Q}$) when the inputs are $\bar{S} = 1$ and $\bar{R} = 0$. The input condition to be avoided in the NAND gate RS latch is $\bar{R} = 0$ and $\bar{S} = 0$ simultaneously. Just as R = 1 and S = 1 in the NOR gate version of the latch, so $\bar{R} = 0$ and $\bar{S} = 0$ in the NAND gate version, creating an undefined output state.

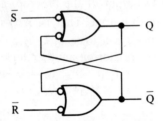

Note: The symbol in this figure is an alternate for the NAND gate.

FIGURE 1-20 Another version of the basic RS latch.

1.15 Race Condition

The most significant reason for avoiding the $\bar{R} = 0$ and $\bar{S} = 0$ input state in the NAND gate latch is the possibility of what is called a *race condition*. The race condition can be created by the following sequence of input signals. Assume that both inputs of the RS latch shown in Figure 1-17 are at logic 0 (and therefore the outputs are both at logic 1). If both inputs are simultaneously raised to a logic 1 state, both NAND gates have two logic 1 inputs, and they must change their output states to logic 0. But the logic 0 outputs remove one logic 1 from each NAND gate, and now both gates must swing back again to logic 1 outputs. These outputs establish two logic 1 inputs to each NAND gate, and we are back where we started—the outputs switch again to logic 0.

As long as the two gates change state at exactly the same time, this sequence continues indefinitely; both gates continuously switch back and forth between logic 0 and logic 1 outputs as fast as they can. Instead of continuing indefinitely, however, one of the gates will switch from one state to the other just a little faster than the other and cause the flip-flop to latch. The gates literally race each other to change states, and the gate that changes its output first prevents the other gate from changing. In this situation, the final outcome cannot be predicted, and this sequence of input states should therefore be avoided. The NOR gate in Figure 1-18 can also be trapped in a race condition if two logic 1 inputs are followed by two logic 0 inputs.

1.16 Other Flip-Flops

Because of the limitations of the RS latch (such as the race condition), other types of flip-flops are needed. There are several versions of a D flip-flop, a J-K flip-flop, and a T flip-flop, but all are based on the RS latch. Each uses the basic RS latch and modifies it in a different manner, so that each displays its own peculiar characteristics and serves a slightly different purpose in logic circuits.

1.17 Clock, Preset, and Clear Signals

The clock signal is a particularly important signal. With reference to Figure 1-21, two gates are connected to the inputs of the latch, and a clock signal is connected so that it can *enable* or *disable* both gates simultaneously. These gates prevent the R and S signals from causing a change in the state of the flip-flop while the clock is low. When the clock is raised to a logic 1, any logic-1 signal on the R or S outputs is gated in (allowed to enter). Then the clock can go to logic 0 again to disable the input gates and keep other signals from entering the latch. The clock signal thus creates what may be called a *window*. Unless this window is open, the state of the flip-flop cannot be changed by the R and S inputs.

Consider how such a window may be useful in a calculator circuit that uses several flip-flops for the storage of numbers. To perform addition, a number is first entered by depressing keys on the calculator keyboard. A clock signal, connected to all flip-flops, is then raised to a

20 · Decision-Making Elements

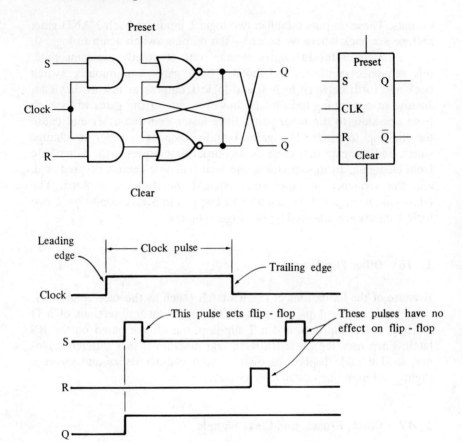

FIGURE 1-21 Operation of a clocked RS latch.

logic-1 level, the numbers are gated into the flip-flops, and the flip-flops are disabled again by placing the clock in the logic-0 state. After that, the clock signal is maintained at logic 0, so that the flip-flops are prevented from receiving additional numbers while calculations are being carried out (the numbers are being processed) by other circuits. Thus, a clock signal can be used to clock or gate data into both inputs of the RS latch.

Another fundamental purpose of a clock is to synchronize. In the foregoing calculator example, storage of data required several flip-flops. The same clock signal was used to enable and disable the data to all flip-flops, thus entering the data into flip-flops synchronously. *Preset* and *clear* are inputs used to set or reset a flip-flop without involving the data and clock inputs. In other words, preset and clear can be used to set and reset the flip-flop, when the clock signal is low. Therefore, it is said that the preset and clear inputs are used to set and reset the flip-flop asynchronously.

To understand how the preset and clear inputs are used, consider that it is not known whether any given flip-flop is in the 1 or 0 state (set or reset) when power is first applied to a circuit. The preset or clear inputs can be used to initialize each flip-flop to a known state and thus ensure that all following logic sequences will proceed correctly. This operation is very similar to clearing a calculator before starting a new calculation.

2 •

COMPUTER AND CALCULATOR ORGANIZATION

2.1 General Considerations

A complete digital computer has a basic architecture, as exemplified in Figure 2-1. Preliminary trouble analysis often involves running specialized programs for evaluation of the computer output. This evaluation serves to localize the fault to a specific area in many cases. In other situations, supplementary tests may be required to close in on the defective subsection. Referring to Figure 2-1, the control is that part of the digital computer that determines the execution and interpretation of instructions in proper sequence, including the decoding of each instruction and the application of the proper signals to the arithmetic unit and other registers in accordance with the decoded information. Frequently, the control denotes one or more of the components in any mechanism responsible for interpreting and carrying out manually initiated directions.

A memory is a storage unit. It is any device into which a unit of information can be copied, which will hold this information, and from which the information can be obtained at a later time. An arithmetic logic unit is that portion of the hardware of a computer in which arithmetic and logical operations are performed. The arithmetic unit generally consists of an accumulator and some special registers for the storage of operands and results, supplemented by shifting and se-

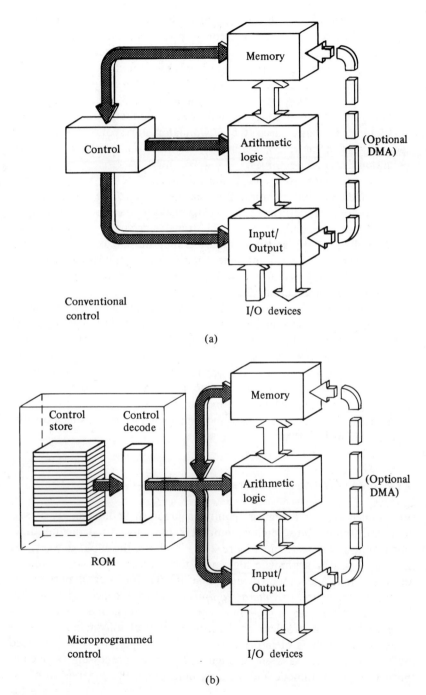

FIGURE 2-1 Basic computer architecture. (a) Standard arrangement; and (b) ROM controlled arrangement. (*Courtesy, Hewlett-Packard*)

quencing circuitry for performing multiplication, division, and other desired operations. An accumulator is a part of the arithmetic unit that may be used for intermediate storage to form algebraic sums and for other intermediate operations. An accumulator may denote a zero-access register and associated equipment in the arithmetic unit in which are formed sums and other arithmetical and logical results. An accumulator may also denote a unit where numbers are totaled (accumulated). In many cases, the accumulator stores one quantity, and upon receipt of any second quantity, it forms and stores the sum of the first and second quantities.

An *operand* is any one of the quantities entering into or arising from a mathematical operation. Zero access refers to the capability of a device to transfer data in or out of a location without undue delays, i.e., owing to other units of data. This transfer occurs in a parallel fashion (simultaneously) and not serially. Input–output, commonly called I/O, is a general term for equipment used to communicate with a computer. It is a process of transmitting information from an external source to the computer, or from the computer to an external source. *Data management accessories* refers to all routines that give access to data, enforce storage conventions, and regulate the use of each I/O device. A routine is a sequence of machine instructions that carry out a well-defined function. Conventions are standard procedures in programs and systems analysis. Microprogramming is the technique of using a special set of computer instructions that consist only of basic elemental operations, which the programmer can combine into higher-level instructions as he chooses and can then program using only the higher-level instructions.

2.2 Memory and ALU Operation

Preliminary trouble diagnosis requires a practical understanding of system functioning. Most digital computers have five fundamental functional units, as depicted in Figure 2-2. These are the input devices, the memory unit, the arithmetic logic unit, the control unit, and the output devices. Input and output devices that are not an integral part of the computer are called *peripheral equipment*. *Data* is a general term that is used to denote any or all facts, numbers, letters, and symbols that refer to a task or problem or describe numerical information. It connotes basic elements of information that can be processed or produced by the computer. As seen in Figure 2-2, data processing is accomplished by the arithmetic and memory units under the supervision of the

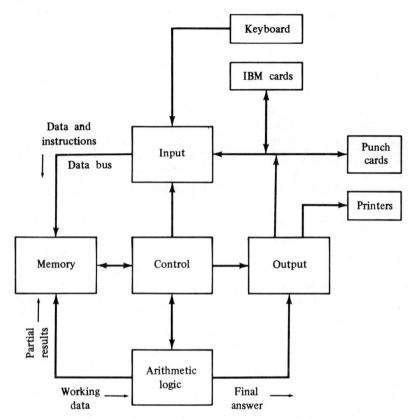

FIGURE 2-2 Functional block diagram of a basic digital computer.

control unit. Data enters through the input devices and exits through the output devices. Data input is ordinarily in the form of digital bits. As noted previously, information units in a computer system are groups of bits called words. *Data words* contain numerical information that is to be processed. *Order* or *instruction words* stipulate the manner in which the data is to be processed. Data words and instruction words are channeled through the computer system via the *information bus,* which may have one or several channels for data flow. This flow may be serial, or it may be parallel.

Information from the programmer is applied to the input section of the computer in various forms. The data might be provided by a teletypewriter, from a telephone line, from a written statement of a problem, and so on. In any case, the information must be translated into binary *machine language* (a program) that can be accepted by the input section. A program is a plan for the automatic solution of a problem. A complete program includes plans for the transcription of data, coding for the computer, and plans for assimilation of the result into the system. The list of coded instructions is called a *routine*. To *transcribe* means to copy, with or without translating, from one external storage medium to another. Often, the data are typed on a keyboard that punches paper tape or cards. Or, the information may be encoded and transferred to magnetic tape for very rapid processing. Data and instructions are fed into the computer input section from which they are entered into storage (in the memory unit). A typical punched card is shown in Figure 2-3. Examples of punched tapes are seen in Figure 2-4.

Digital words were noted previously. A word is a set of characters that occupies one storage location and is treated by the computer circuits as a unit; it is transported as a unit. A word is treated by the control unit as an instruction and is treated by the arithmetic unit as a quantity. Word lengths may be fixed or variable, depending on the design of the particular computer. It is instructive to observe a typical binary *data word* that is to be stored by a register in the memory section. A binary word comprising 34 bits is shown in Figure 2-5. Note that 32 bits are utilized for the binary coded decimal digits, plus the algebraic sign (+ or −), plus the parity bit. Thus, a total of 34 bits is employed in this example. This word happens to correspond to 14 pulses. Of course, a data word that has different decimal digits would correspond to a different number of pulses.

Next, consider the binary *instruction word* shown in Figure 2-6. A total of 34 bits is utilized in this example. The 34th bit is a parity check, and the 28th through 33rd bits are an *operation code*. Twenty-seven bits are assigned to *addresses*. An operation code consists of the symbols that designate a basic computer operation that is to be performed. It is the part of an instruction that designates the operation of arithmetic or of logic or the transfer to be performed. As noted previously, an address is a number that identifies a register, location, or unit where information is stored. It is the operand part of an instruction which extracts a specific piece of information from the memory or puts information in the memory. Figure 2-7 depicts the basic structure of a core memory. One polarity of current flow magnetizes a core in one direction; the opposite polarity of current flow magnetizes the core in the other direction. These two states are represented by binary 0 and 1. Note that an individual core can be addressed by energizing a particular pair of wires, such as X_1 and Y_2.

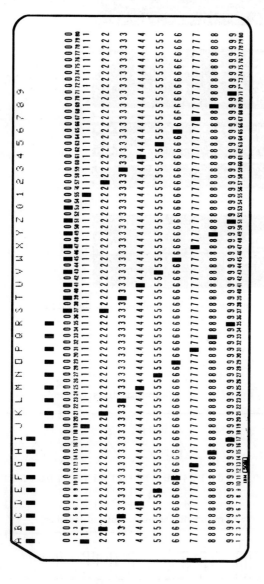

FIGURE 2-3 A typical punched card. (*Courtesy, IBM*)

28 · Computer and Calculator Organization

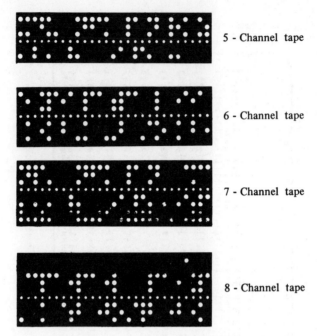

FIGURE 2-4 Examples of punched tapes.

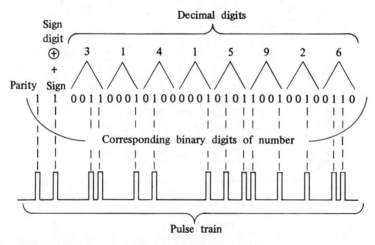

FIGURE 2-5 A binary data word comprising 34 bits.

2.2 Memory and ALU Operation · 29

Parity	OP code	Addresses
(34)	010010	100110111001000101110000101
	(28 to 33)	(1 to 27)
1 bit	6 bits	27 bits
		Total of 34 bits

FIGURE 2-6 A binary instruction word comprising 34 bits.

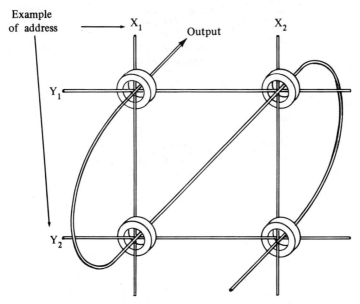

FIGURE 2-7 Basic structure of a core memory.

It is instructive to note the essentials of arithmetic-unit operation. All arithmetical operations are accomplished in terms of addition or of extracting a value stored in a memory. To add two binary numbers, the following rules are followed:

$0 + 0 = 0$
$1 + 0 = 1$
$0 + 1 = 1$
$1 + 1 = 1$ and carry 1

A basic logic circuit that is used to start the operation of addition is the XOR gate, exemplified in Figure 2-8. It is only a parity checker,

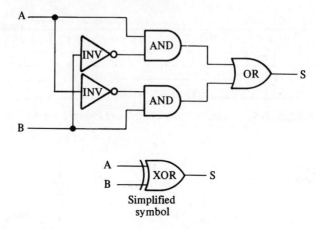

FIGURE 2-8 An XOR circuit, the basic binary addition configuration. (*Courtesy, Hewlett-Packard*)

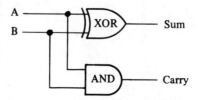

FIGURE 2-9 Half-adder configuration.

because it does not generate a carry 1. Therefore, the XOR gate is elaborated by including an AND gate, as depicted in Figure 2-9. This arrangement is called a *half adder*. It can add two binary bits. A separate adder circuit is required to add each pair of bits, and the addition can be done in parallel. When adding in parallel, however, each bit position affects the addition to the left of that bit position.

```
    1  0 1 1  ← C   (carries)
       1 0 0 1 ← A  (augend)
     + 1 0 1 1 ← B  (addend)
    1  0 1 0 0 ← S  (sum)
```

Accordingly, in the computer system, each adder must be connected to process three inputs: A, B, and the carry C. In turn, a *full*

2.2 Memory and ALU Operation · 31

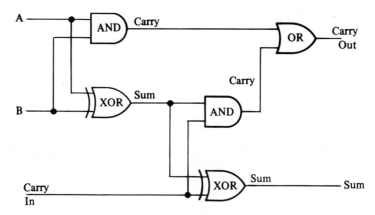

FIGURE 2-10   A full-adder configuration.

adder circuit has the configuration shown in Figure 2-10. A full adder can be constructed in many different ways, but it always performs the same function in that it adds two bits and a previous entry to generate a sum as well as a carry. By using special algorithms, a full adder can also perform subtraction, multiplication, and division. An algorithm is a fixed step-by-step procedure for accomplishing a given result.

A parallel adder consists of several full-adder stages that are interconnected so that the carry output from one stage becomes the carry input to the next stage, as shown in Figure 2-11. Thus, a four-stage parallel adder will handle the propagation of all carries and can be used to add any two 4-bit binary numbers. There are several significant features in a parallel adder that should be understood by the

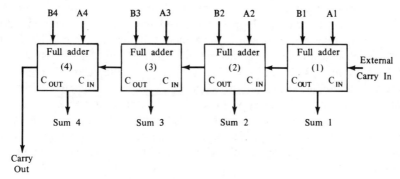

FIGURE 2-11   A parallel 4-bit adder configuration.

digital troubleshooter. A so-called parallel adder operates in a sequential manner. For example, if the following numbers are added,

$$\begin{array}{r} 1111 \leftarrow C \quad \text{(carries)} \\ 0111 \leftarrow A \quad \text{(augend)} \\ +1001 \leftarrow B \quad \text{(addend)} \\ \hline 10000 \leftarrow S \quad \text{(sum)} \end{array}$$

it will be noted that a carry is generated by each stage of the addition. The first adder must complete the addition of A1 and B1 in order to generate the carry C1 input to the second adder. In turn, the second adder cannot correctly perform its part of the addition sequence until the first adder has completed adding and generating carry C1. Similarly, the third adder must wait for the second adder to generate carry C2, and so on.

Observe that the carries pass sequentially through all stages of the parallel adder and that the last sum output is not correct until the last carry is generated. This carry propagation is called *ripple carry*. It follows that every stage of the adder in Figure 2-11 performs an addition as soon as its A and B inputs are present, but the outputs of any stage may not be correct until a carry input has been processed. Thus, an interim incorrect sum output may be generated that cannot be allowed to enter into other circuits of the system. For this reason, parallel adders often have sum output gates that are disabled until the adder has had sufficient time to generate the correct sum signals.

A basic serial adder is arranged around a single full adder circuit that adds each pair of bits sequentially. The auxiliary circuitry that comprises the remainder of the serial adder is arranged so that the augend word is entered from an external memory storage into the A shift register depicted in Figure 2-12, and the addend is entered into the B shift register; both words have their LSB on the right. This LSB from each register is shifted into the full adder. Then, the sum is shifted into the sum register, and the next two bits from the A and B registers are shifted to the adder on the same clock pulse. If the first addition produces a carry, the carry is stored in FF1 and becomes an input to the full adder during the next addition. A pair of binary words of any length can be added in a series of such additions, starting with the LSB and ending with the MSB, clocking all shift registers with the same clock signal.

Several modifications may be encountered to this basic serial

2.2 Memory and ALU Operation · 33

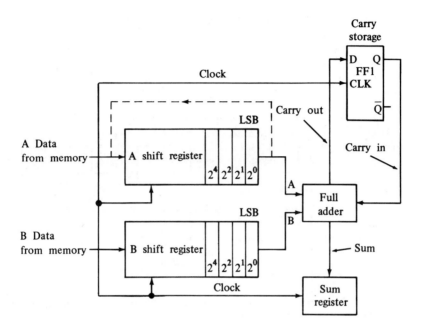

FIGURE 2-12 Serial adder configuration. (*Courtesy, Hewlett-Packard*)

adder in different applications. If it is desired to add several different numbers in sequence to the same augend, the augend is shifted back into the A register in a circular manner, as indicated by the broken line in Figure 2-12. At the end of any addition, the augend is back in its original position in the A register, ready for the next addition. If it is desired to obtain a sum of more than two numbers, the first two numbers can be added, and then the third number is added to the sum of the first two, and so on. For this purpose, the sum is shifted back into register A, which is then called an *accumulator,* as indicated in Figure 2-13.

There is a commonly used method of subtraction of binary numbers in computer systems that is based on *adding* the complement of one number to another number instead of subtracting one number from another. This permits simplification of the logic circuitry. The 1's complement of a binary number is formed by changing the 1's to 0's, and changing the 0's to 1's. The following example shows how subtraction is performed directly and by the addition of a 1's complement:

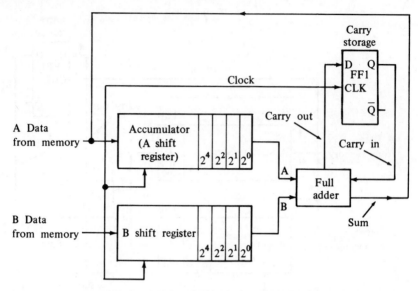

FIGURE 2-13 Serial adder with accumulator. (*Courtesy, Hewlett-Packard*)

DIRECT SUBTRACTION			SUBTRACTION BY THE ADDITION OF 1'S COMPLEMENT

Decimal Binary

```
    9     1001    Minuend          1001
   -3     0011    Subtrahend       1100    complement the subtrahend
   ──     ────                    ──────
    6     0110                   1 0101    add to the minuend
                                 └──→1     add end-around carry to LSB
                                   ────
                                   0110    difference
```

Note that the basic principle in the foregoing example is to complement the subtrahend and add it to the minuend. This is the procedure for any subtraction by complements—the end-around carry is explained in this section. Subtraction by addition of a complement is widely used in preference to direct subtraction because it eliminates the need for separate subtracter circuits—only adder circuits and the auxiliary logic to complement numbers and to handle end-around carry are needed. To implement direct binary subtraction, a *half subtracter* and a *full subtracter* that correspond to half and full adders may be employed. Figure 2-14 shows one version of a full subtracter, but just as in adder circuits, various types of gates can be used to provide the

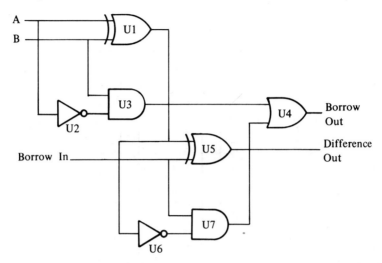

FIGURE 2-14 A full-subtracter configuration. (*Courtesy, Hewlett-Packard*)

same result. Adder and subtracter circuits are so similar that it is also possible to construct a combined adder/subtracter circuit, with an ability to switch from one operation to the other by means of a single control signal.

Full subtracters are not encountered as frequently as other types of subtracters in computer systems. Most calculators and computers perform subtraction by the addition of a complement, using only full adder circuits for all addition and subtraction operations. This design reduces the amount of logic required in any one circuit, increases the speed of operation, and simplifies programming. Because subtraction can be performed by the addition of a complement, the distinction between addition and subtraction in typical computer logic becomes comparatively small. A typical computer simply recognizes positive and negative numbers and adds these numbers. To distinguish between positive and negative numbers, a + sign is indicated by 0 in the left-most digit of the number if it is positive. If the number is negative, a − sign is indicated by a 1 in the left-most digit of the number. The computer circuitry responds to these digit positions as if they were algebraic signs. Three general methods of binary number representation are employed in computer systems. The method that is employed affects how negative numbers are stored in the memory unit and, in turn, affects the way that addition and subtraction are performed. These three methods are:

EXAMPLES	ALGORITHMS
	Two Numbers Of Same Sign
(Positive) (Negative) +13 0.01101 −13 1.10010 +11 0.01011 −11 1.10100 +24 0.11000 −24 1.00110 ↰1 1.00111	To augend magnitude add addend magnitude; retain existing sign To 1's complement of augend magnitude add 1's complement of addend magnitude; add end-around carry; retain existing sign (answer in 1's complement form)
	Two Numbers Of Opposite Sign
−13 1.10010 +10 0.01011 +11 0.01011 −13 1.10010 −2 1.11101 −2 1.11101	To magnitude of augend (in 1's complement form, if negative) add magnitude of addend (in 1's complement form, if negative).
	Two Numbers Of Opposite Sign
+13 0.01101 −11 1.10100 −11 1.10100 +13 0.01101 +2 0.00001 ↰1 +2 0.00001 ↰1 0.00010 0.00010	To magnitude of augend (in 1's complement form, if negative) add magnitude of addend (in 1's complement form, if negative); add end-around carry.

Note: All examples shown in the table are defined to be additions. Subtraction is defined to be for example, +13 − (−11). Subtraction requires the complementing of the subtrahend in a true/complement circuit, such as shown in Figures 2-17 and 2-19

FIGURE 2-15   Addition and subtraction by 1's complements. (*Courtesy, Hewlett-Packard*)

1. *1's Complement.* Positive binary numbers are stored in the memory as a binary code with a 0 preceding the number to indicate a positive sign. Negative numbers are stored in complement form with a 1 preceding the number to indicate a negative sign. Figure 2-15 shows examples of the addition and subtraction processes that are performed. Typical logic for a 1's complement adder/subtracter is depicted in Figure 2-16.

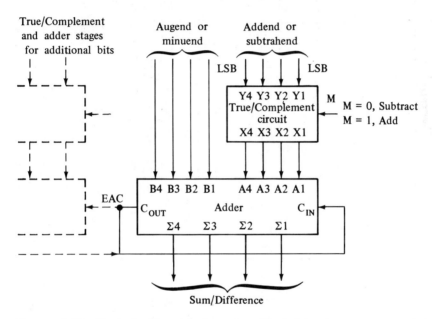

FIGURE 2-16 Typical 1's complement adder/subtracter arrangement. (*Courtesy, Hewlett-Packard*)

2. *2's Complement.* Positive binary numbers are stored in the memory in binary code with a 0 in the sign position. Negative numbers are stored in 2's complement form; a sign bit of 1 indicates the negative sign. The 2's complement is derived by adding 1 to the 1's complement. Addition takes place directly and subtraction requires complementing the subtrahend before the addition takes place. Unlike operations with 1's complement numbers, 2's complement operations never generate an end-around carry. Examples of 2's complement method are shown in Figure 2-17. Typical logic for a 2's complement adder/subtracter is depicted in Figure 2-18.
3. *Sign and Magnitude.* Both positive and negative numbers are

EXAMPLES	ALGORITHMS
Two Numbers Of Same Sign	
(Positive) (Negative) +13 0.01101 −13 1.10011 +11 0.01011 −11 1.10101 +24 0.11000 −24 1.01000	To augend magnitude add addend magnitude; retain existing sign. To 2's complement of augend magnitude add 2's complement of addend magnitude, neglect last carry; retain existing sign.
Two Numbers Of Opposite Sign - Augend Magnitude Larger	
+13 0.01101 −13 1.10011 −11 1.10101 +11 0.01011 +2 0.00010 −2 1.11110	To magnitude of augend add 2's complement of addend; add sign bits also, but neglect carry from sign bits. To 2's complement of augend magnitude add magnitude of addend; add sign bits also, but neglect carry from sign bits.
Two Numbers Of Opposite Sign - Addend Magnitude Larger (Or Equal)	
−11 1.10101 +11 0.01011 +13 0.01101 −13 1.10011 +2 0.00010 −2 1.11110	To 2's complement of augend magnitude add magnitude of addend; add sign bits also, but neglect carry from sign bits. To magnitude of augend add 2's complement of addend; add in sign bits also, but neglect carry from sign bits.

Note: See footnote to Figure 2-15 for a definition of addition and subtraction.

FIGURE 2-17 Addition and subtraction by 2's complements. (*Courtesy, Hewlett-Packard*)

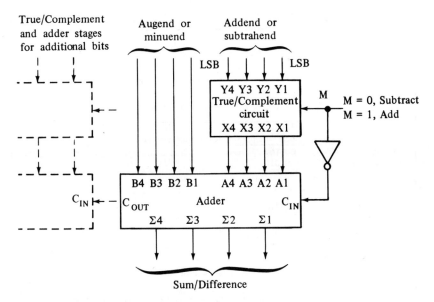

FIGURE 2-18 Typical 2's complement adder/subtracter. (*Courtesy, Hewlett-Packard*)

stored in the computer memory in uncomplemented form. The distinction between positive and negative numbers is the sign bit. Unlike 1's complement and 2's complement, arithmetic addition and subtraction operations in sign and magnitude format are not the same for all combinations of numbers. In other words, to know which (if any) of the numbers must be complemented in an addition or subtraction operation, and to know the sign of the sum or difference, it is necessary to know the signs of the numbers involved as well as which number is larger. Examples of the sign-and-magnitude method are shown in Figure 2-19. Typical logic for a sign-and-magnitude adder/subtracter is depicted in Figure 2-20.

Next, it is important for the digital troubleshooter to understand the essentials of binary multiplication in computer systems. Four basic rules, or algorithms, of binary multiplication are as follows:

$0 \times 0 = 0$
$0 \times 1 = 0$
$1 \times 0 = 0$
$1 \times 1 = 1$

40 · Computer and Calculator Organization

EXAMPLES	ALGORITHMS
Two Numbers Of Same Sign	
(Positive) (Negative) +13 0.01101 −13 1.01101 +11 0.01011 −11 1.01011 +24 0.11000 −24 1.11000	To magnitude of augend add magnitude of addend retain existing sign.
Two Numbers Of Opposite Sign - Augend Magnitude Larger	

+13   0.01101   −13   1.01101
 −11   1.01011   +11   0.01011
 +2   −2
 	
     01101   01101
     10100   10100
     ⌐00001   ⌐00001
     ⌊→1   ⌊→1
     00010   00010
     0.00010   1.00010
	To magnitude of augend add 1's complement of addend magnitude;
	
	 end-around carry
	
	 retain sign of augend.

Two Numbers Of Opposite Sign - Addend Magnitude Larger (Or Equal)	

+11   0.01011   −11   1.01011
 −13   1.01101   +13   0.01101
 −2   +2
 	
     01011   01011
     10010   10010
     11101   11101
     1.00010   0.00010
	To magnitude of augend add 1's complement of addend magnitude
	
	 (there is no end-around carry);
	
	 complement result and retain sign of addend.

Note: See footnote to Figure 2-15 for a definition of addition and subtraction.

FIGURE 2-19 Addition and subtraction by sign and magnitude. (*Courtesy, Hewlett-Packard*)

These fundamental algorithms are the same as in the decimal number system. However, in binary multiplication, they are used to develop more comprehensive algorithms for processing large numbers and fractions as well as positive and negative numbers. At this point, two simple and familiar properties of multiplication should be considered. First, the process of multiplication is repeated addition. In other words, 4 × 10 denotes adding 10 to itself four times. This principle is equally true in the binary and decimal systems. Second, as has been noted previously, shifting a binary number to its left by one posi-

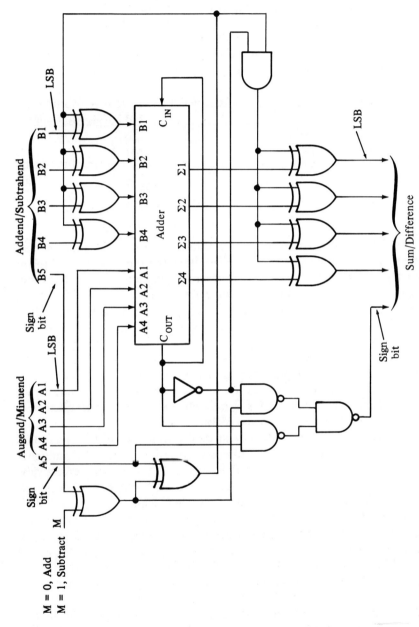

FIGURE 2-20 Typical sign-and-magnitude adder/subtracter. (*Courtesy, Hewlett-Packard*)

tion is equivalent to multiplication by 2. Note that in the decimal system, a similar shift is equivalent to multiplication by 10. The foregoing principles of repeated addition and multiplication by shifting are both used in algorithms for binary multiplication (as well as for division). In the following example, a binary multiplication is shown in detail, to illustrate how an arithmetic circuit performs multiplication:

```
(Decimal)   13           (Binary)    1101  ←— Multiplicand
          × 11                       1011  ←— Multiplier
            13                       1101 ⎫
            13                       1101 ⎪
           143                       0000 ⎬ Partial Products
                                     1101 ⎭
                                  10001111 ←— Product
```

Each of the partial products results from a multiplication by 1 or by 0. That is, each partial product is either the same as the multiplier, or it is 0. The final product is the result of addition of all partial products, each succeeding partial product being shifted to the left by one position. In the foregoing example, all partial products were added simultaneously. However, the final product could have been obtained by adding the first two partial products, then adding the third to the sum, and so on, always shifting each subsequent partial product left by one position. In consequence, it can be stated that, in its simplest form, binary multiplication is a series of additions of the multiplicand, with a shift after each addition. Thus, a multiplication can be performed by a circuit that comprises adders and registers.

Next, observe how the circuit depicted in Figure 2-21 multiplies the binary numbers from the foregoing example. First, the multiplier and the multiplicand are entered into B and A registers, respectively, and the accumulator is cleared. Next, the add/shift control logic examines the right-most digit of the multiplier; inasmuch as it is a 1, it instructs the adder to add the multiplicand to the contents of the accumulator (this first step represents the storage of the first partial product 1101 from the foregoing example). After the addition, the add/shift control logic shifts both the B register and the accumulator right by one place. Now, the right-hand digit of the multiplier is examined again, and, because it is a 1, the multiplicand is added to the accumulator a second time (representing the addition of the second partial product 1101). This is followed by another shift and examination of the right-most multiplier digit. This time, however, the multiplier

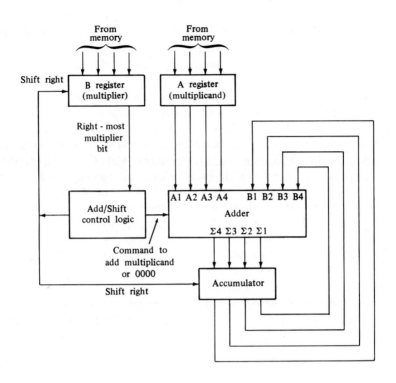

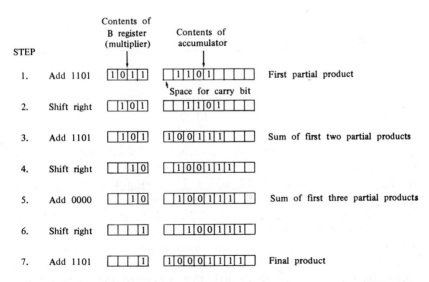

FIGURE 2-21 Example of a binary multiplier. (*Courtesy, Hewlett-Packard*)

digit is 0; therefore, nothing is added to the accumulator (representing the third partial product 0000)—it is simply shifted right. The last step is another addition of the multiplicand and the final product is now stored in the accumulator.

Because division is the opposite of multiplication, it can be performed by repeated subtraction. Besides the four basic functions of addition, subtraction, multiplication, and division, calculators and computers also perform various other mathematical operations. Such operations as the derivation of square roots and the computation of logarithms and trigonometric functions are all executed with the same logic circuits that have been described. Various algorithms are employed. Often, these algorithms are iterative processes—a step or a sequence of steps is repeated over and over again until the desired accuracy of the answer is obtained. Note that the mathematical operations that have been described were performed in straight binary code. This is almost always true of addition and subtraction in computers. On the other hand, in calculators, the binary coded decimal (BCD) code is often employed. A BCD adder or subtracter is not greatly different from straight binary adders and subtracters. They add or subtract by means of the same algorithms, with the data processed in BCD format.

It is instructive for the troubleshooter to observe how two numbers are added in the 8421 BCD code:

(Decimal)	HUNDREDS	TENS	ONES	← BCD Digit Groups
1				
386	0011	1000	0110	
243	0100	0100	0011	
629	0101	1100	1001	

In decimal form, the addition, including the carrying of a 1, is familiar. In BCD form, the addition of each of the digit groups separately results in three numbers: 5, 12, and 9. If the sum were read directly, it would be written as 5(12)9, which is, of course, incorrect. Note that the correct method requires that the 2 be written in the ten's position with the 1 carried to the hundred's position to change the 5 to a 6— just as it is done in decimal addition. When a number is added in BCD form by logic circuits, special auxiliary logic is necessary to detect the carry and to correct the BCD digit that generated the carry. In the previous example, in addition to the carry detection and its propagation to the hundred's group, the ten's group must be corrected to output 2 in binary and not 12 (it must be changed from 1100 to 0010). Figure 2-22 shows a BCD adder circuit that can generate the carry and correct the adder output of one BCD digit (4 bits).

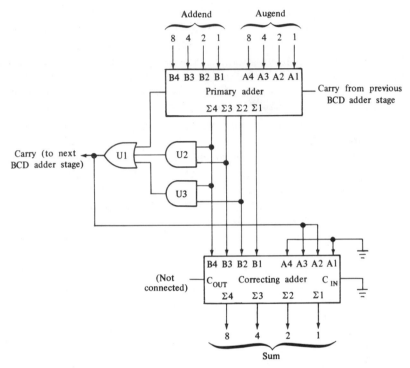

FIGURE 2-22 BCD adder stage configuration. (*Courtesy, Hewlett-Packard*)

Note that even though a 4-bit binary full adder can add two numbers whose sum is as large as 15, without generating a carry, both a carry and the accompanying correction are necessary any time that the sum is 10 or greater (just as a carry is done when decimal numbers are added). In the foregoing example, only the ten's group generated a carry and necessitated a correction, but in the general case, any one of the digit groups can exceed 9 during an addition. To add, for example, two three-digit BCD numbers together, three parallel adder stages are needed, each with the ability to generate a carry and to correct its output when necessary. To generate the carry, a detection circuit must be present in each adder stage to output a carry if the sum is greater than

46 · Computer and Calculator Organization

9. This detector consists of a few decoder gates (U1 through U3 in Figure 2-22) that detect 10, 11, 12, 13, 14, or 15, or the binary carry for sums of 16 through 19. The two types of carries are ORed together for generating the carry to the adder stage.

To correct the sum output of any stage that has generated a carry, it turns out that if 6 (binary 0110) is added to any uncorrected sum, the result will be correct. (Since 0110 is the 2's complement of 1010, adding 1010 is the same as subtracting 1010, or decimal 10.) The correction is made in another adder where a 0110 is added to the uncorrected sum. If no carry is generated, the second adder simply adds 0000 to the sum output of the first adder. Figure 2-22 illustrates how the presence of a carry signal at the output of U1 causes a 0110 to be added in the correcting adder and how the absence of a carry causes a 0000 to be added. The adder stage depicted in Figure 2-22 can be connected in parallel with several identical stages to form a multidigit BCD adder. The 4-bit adders in each stage can be either a simple parallel type or an elaborated type.

Although the more basic principles of system operation have been explained, the digital troubleshooter will often find it necessary to consult the operating and servicing manuals for the particular computer with which he may be concerned. An analysis of incorrect outputs requires familiarity with the system block diagrams and flow charts. After this knowledge has been acquired, it is good practice to follow an established generalized troubleshooting procedure, as explained in the following chapters.

3 •
BASIC DIGITAL TEST INSTRUMENTS

3.1 Troubleshooting With a Logic Probe

Digital technicians often state that a logic probe is the most important single unit of digital test equipment. A probe such as the one illustrated in Figure 3-1 is used to test the state of logic to determine whether a terminal is logic-high, logic-low, or faulty. A faulty indication is produced by an open circuit or by subnormal-amplitude pulses. Logic probes are particularly useful in preliminary trouble analysis. After a defective area has been located with a logic probe, an oscilloscope is generally utilized to determine the details of the defect or failure. The logic probe shown in Figure 3-1 contains two lights, red and green, in the nose of the probe. The red light indicates logic 1 (logic-high) in the range from 2.15 to 5 volts. The green light indicates 0 (logic-low) in the range from 0 to 0.7 volt. Probe indications corresponding to various logic states are as follows:

PROBE RESPONSE	LOGIC STATE
Steady red light	Steady logic-high state
Steady green light	Steady logic-low state
Blinking red and green at full intensity	Pulse trains

No light   Abnormal state between logic-high and logic-low

No light alternatively   Open circuit or resistance greater than 10 ohms

Both red and green lit   Abnormal input voltage (greater than 6 volts)

Blinking red light   Alternating between high state and indeterminate state

Blinking green light   Alternating between low state and indeterminate state

Green, red, then green   Single positive pulse

Red, green, then red   Single negative pulse

If the pulse repetition rate is less than 5 Hz, the lights will blink and follow the pulse rise and fall. The input impedance of the digital logic probe is 7.5 k paralleled by 6 pF; this impedance is sufficiently high that the circuit under test is not disturbed. When testing TTL circuitry that operates at high repetition rates, a short ground lead should be used with the logic probe. A short ground lead avoids the possibility of overshoot and ringing in the processed pulses, which could cause false probe indications. In most situations, power is applied to the logic probe from the circuit under test via clip leads attached to the probe. A +5 volt source is utilized.

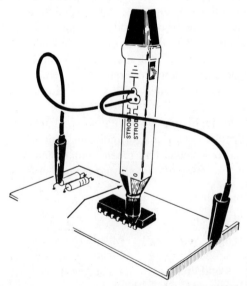

FIGURE 3-1   Checking IC's with a digital logic probe. (*Courtesy, Tektronix, Inc.*)

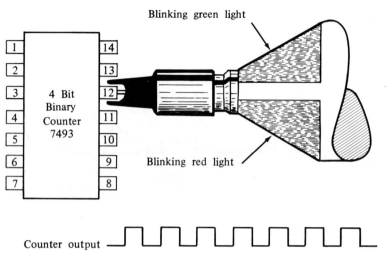

FIGURE 3-2 Logic probe applied for checking counter output. (*Courtesy, Tektronix, Inc.*)

Referring to Figure 3-2, the logic probe is being applied to check the output from a binary counter. In this example, the red and green lights blink on and off at a 5-Hz rate, indicating that the logic level is going high and low alternately. Next, Figure 3-3 shows the logic probe being applied to check the inputs and outputs of a NAND gate. Thus, when the two inputs are logic-low, the output is logic-high. In turn, when pin 8 is checked, the red light normally turns on. Again, when the input pins 9 and 10 are checked, the green light normally turns on, indicating that the pins are logic-low. In various situations, the troubleshooter needs to know whether a single pulse has passed. As an illustration, a certain pulse might occur only once in several minutes. For this type of test, the probe provides a storage function that captures a single pulse with a width as small as 10 nsec. When the probe is switched to its storage mode, it can be secured to a test point as depicted in Figure 3-4. Thereby, the test point can be monitored.

With reference to Figure 3-4, the green light normally turns on and remains on when pin 8 is checked, indicating a logic 0. Then, the red light normally turns on when the four inputs become positive, causing the output pin 8 to go positive, which is a logic 1 state. Even if the probe is disconnected from pin 8, the two lights will remain on. In order to reset the probe, the storage switch is thrown back to its original position. If the storage function is needed again, the storage switch setting is reversed. If desired, the logic probe may be secured to a test

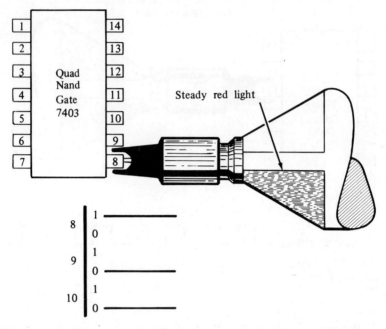

FIGURE 3-3 Logic probe applied for checking a NAND gate. (*Courtesy, Tektronix, Inc.*)

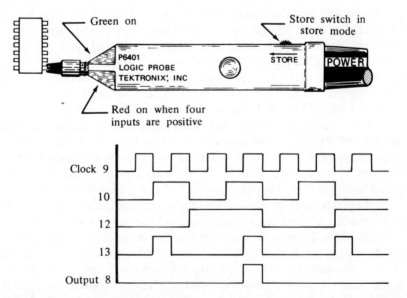

FIGURE 3-4 Logic probe operated in its storage mode. (*Courtesy, Tektronix, Inc.*)

point and left unattended. During the monitoring test, if a pulse occurs, the corresponding light will turn on and will remain on indefinitely. Thus, the probe indication may be checked at intervals to determine whether an expected pulse has occurred.

In various situations, the digital troubleshooter needs to detect the coincidence of two pulses. For this purpose, a strobe input is provided by the logic probe, as seen in Figure 3-5. To check whether two pulses occur at the same time, the strobe input of the probe is connected to the strobe point where coincidence normally occurs. This might be a gate or a strobe terminal. In case the gate or strobe terminal

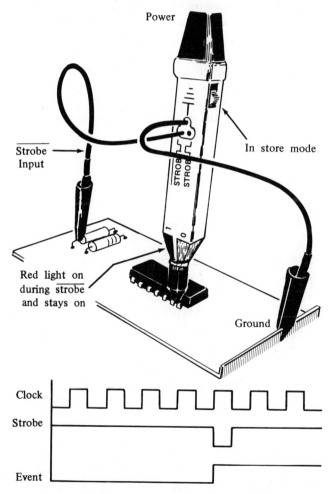

FIGURE 3-5 "Event" checking with a logic probe. (*Courtesy, Tektronix, Inc.*)

52 · Basic Digital Test Instruments

happens to supply a negative pulse, the logic probe is operated on its strobe (strobe NOT 0 function). The terminal in question is then monitored by applying the probe tip, as depicted in Figure 3-5. The probe recognition circuitry is gated off until a strobe pulse appears. However, probe response will occur with the appearance of a strobe pulse. This event will be indicated by the probe lights and will be retained if the storage mode is employed. Both the red and the green lights will turn on.

If the logic probe is applied and the anticipated event does not occur, the fault will usually be traced to a defective component, such as a malfunctioning IC. It may be found that the IC inputs are correct but that the output is incorrect. In most cases, replacement of the defective IC (or other component) will restore the digital system to normal operation. On the other hand, if additional analysis should be required, an oscilloscope can be employed to check the waveshapes and voltage levels for a more detailed trouble analysis. An oscilloscope that is used to check digital pulse waveshapes should have an appreciably faster rise than the rise time of the pulses. Figure 3-6 shows the definition of pulse rise time.

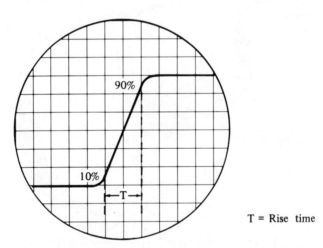

T = Rise time

Figure 3-6 Definition of pulse rise time.

Typical pulse waveforms encountered in digital-equipment troubleshooting are shown in Figure 3-7. Pulse trains, combinations of pulse and rectangular waveforms, and infrequent pulses may be displayed. One characteristic that is common to all is the normal amplitude

3.1   Troubleshooting With a Logic Probe · 53

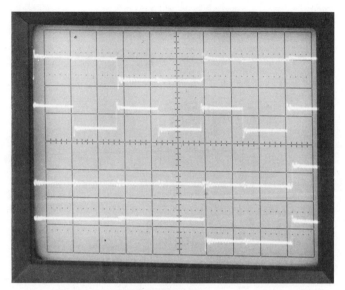

FIGURE 3-7   Typical pulse waveforms encountered in digital-computer troubleshooting. (*Courtesy, Hewlett-Packard*)

(logic-high) level. There is no sharp demarcation between a pulse and a rectangular waveform. In other words, narrow pulses will be called *wide pulses* at some arbitrary point of increasing width. Again, wide pulses will be called *rectangular waveforms* at some arbitrary point of increasing duration. Trouble symptoms in digital circuitry may be caused by pulses that have become too wide or that have been otherwise distorted by defective components or devices. Figure 3-8 pictures the basic forms of rectangular wave distortion. A digital waveform is subject to certain tolerances; this is an involved topic, and expertise in evaluation of waveform tolerances is acquired chiefly by practical experience.

In any digital system, there is a permissible tolerance on pulse distortion, just as there is a permissible tolerance on pulse amplitude. In other words, as the amount of distortion increases, there will arrive a point at which system operation becomes marginal and unpredictably erratic. Further increase in distortion results in a sustained trouble symptom. Practical troubleshooting procedures often require knowledgeable judgments of waveform tolerances. In case of doubt, the troubleshooter can energize a device or subsystem by pulses from a digital pulse generator in order to check the resulting operation when a

54 · Basic Digital Test Instruments

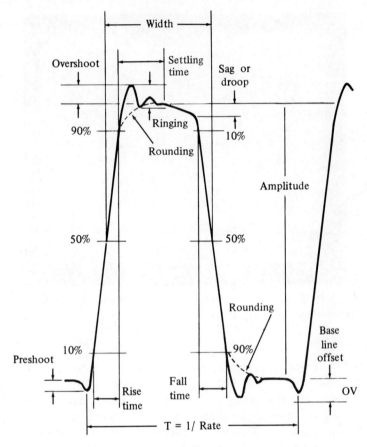

FIGURE 3-8 Basic forms of rectangular-wave distortion.

near-ideal pulse input is applied. Digital-equipment servicing data may specify normal pulse shapes at various points through the system.

3.2 Troubleshooting With the Logic Pulser and Logic Probe

A logic pulser is basically a single-shot pulse generator with a high output-current capability. It is generally used with a logic probe to provide indication of pulse activity, faulty in-circuit IC's, and the static states of all pins, as illustrated in Figure 3-9. The circuit under test can be stepped one pulse at a time while checking the truth tables of the logic

3.2 Troubleshooting With the Logic Pulser and Logic Probe · 55

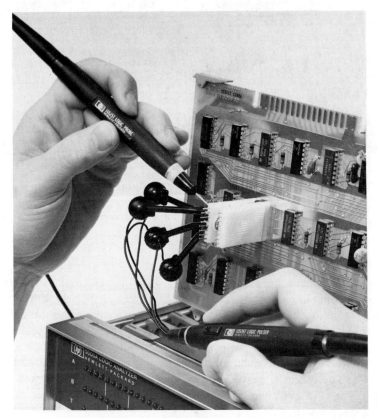

FIGURE 3-9 A logic pulser and logic probe localize faults in logic circuitry. (*Courtesy, Hewlett-Packard*)

packages in order to turn up any defects. That is, a logic pulser provides a convenient means of injecting single pulses, the effects of which are monitored with a logic probe. Note that the operation of the pulser is automatic and that no adjustments are required. The tip of the logic pulser is touched to the circuit under test (Figure 3-10), and the pulse button is pressed. In turn, all circuits connected to the terminal (outputs as well as inputs) are briefly driven to their opposite state. No unsoldering of IC outputs is required.

Note that the troubleshooter does not concern himself with whether the test point is logic-high or logic-low, because high nodes are automatically pulsed low, and low nodes are automatically pulsed high, each time the button is pressed. The pulser illustrated in Figure 3-10

56 · Basic Digital Test Instruments

FIGURE 3-10 Logic pulser in use to troubleshoot digital circuitry. (*Courtesy, Hewlett-Packard*)

will source or sink up to 0.65 ampere, which is sufficient to override IC outputs in either the high or low state. An output pulse width of 0.3 μsec is employed to limit the amount of energy delivered to the device under test, thereby eliminating the possibility of damage from excessive test energy. This type of pulser can be used with either TTL or DTL logic. Circuit action is unaffected by probing until the pulse button is pressed. The pulser is powered from any 5-volt supply and draws less than 25 mA. In most cases, the pulser is connected to the power source of the system under test.

Consider the characteristics of the Hewlett-Packard logic probe illustrated in Figure 3-11. It is employed to trace logic levels and pulses through integrated circuitry to determine whether the point under test is logic-high, low, bad level, open-circuited, or pulsing. This probe has

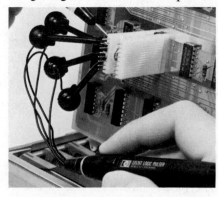

FIGURE 3-11 Logic probe used with the logic pulser shown in Figure 3-10. (*Courtesy, Hewlett-Packard*)

3.2 Troubleshooting With the Logic Pulser and Logic Probe · 57

preset logic levels of 2.0 and 0.8 volts, which correspond to the high and low states of TTL and DTL circuitry. When the probe is touched to a high-level point, a bright band of light appears around the probe tip. When the probe is touched to a low-level point, the light goes out. Open circuits or voltages in the "bad level" region between the preset thresholds produce illumination at half brilliance. Single pulses of 10 nsec or greater widths are made readily visible by stretching up to one-twentieth of a second. The probe lamp flashes on or blinks off depending upon the polarity of the pulse. Pulse trains up to 50-MHz repetition rate cause the lamp to blink off and on at a 10-Hz rate.

Figure 3-12 shows the logic probe in use. The circuit under test can first be operated at normal speed while checking for the presence of key signals, such as clock, reset, start, shift, and transfer pulses. Next, the circuit under test can be stepped one pulse at a time while checking the truth tables of the logic packages in order to turn up any defects. Connectors provided with the logic probe facilitate connection to the required 5-volt supply from either the circuit under test or an auxiliary power supply. A ground clip is provided which may be used to directly connect the probe circuitry to the ground of the circuit under test. A block diagram of this logic-probe circuitry and its response to various inputs are shown in Figure 3-13.

FIGURE 3-12 Logic probe in use. (*Courtesy, Hewlett-Packard*)

58 · Basic Digital Test Instruments

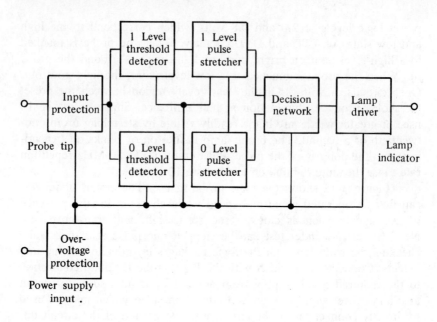

Logic probe block diagram

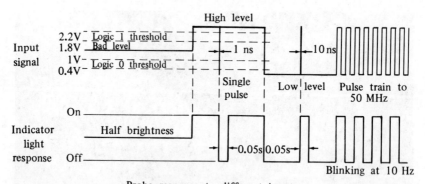

Probe response to different inputs

FIGURE 3-13 Block diagram of the logic-probe circuitry and its response to various inputs. (*Courtesy, Hewlett-Packard*)

Referring to Figure 3-14, operation of this logic probe is as follows. Only the Logic One channel is described, inasmuch as the Logic Zero channel is the same following the threshold detector. In this section, the positive-true logic convention is used:

TRUE = ONE (1) = HIGH (more positive potential)
FALSE = ZERO (0) = LOW (more negative potential)

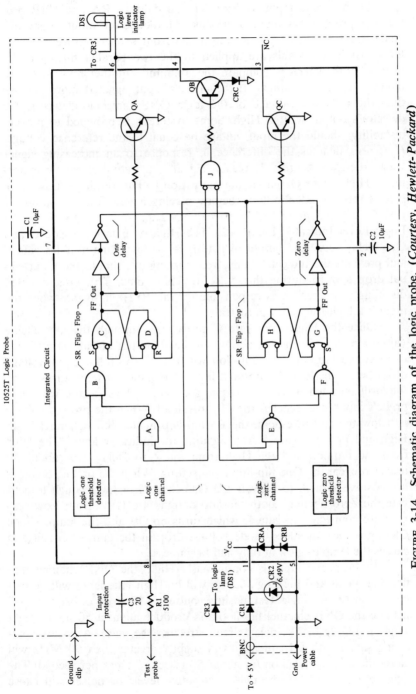

FIGURE 3-14 Schematic diagram of the logic probe. (*Courtesy, Hewlett-Packard*)

59

At the probe input, resistor R1 and diodes CRA and CRB protect the probe from input overloads. At the power input connector, diodes CR1 and CR3 protect the probe from reversed power connections. All input signals are applied to the two parallel threshold detectors. Both detectors compare the amplitude of the input signal to internal reference voltages. If the probe's input signal is more positive than the reference voltage of the logic ONE threshold detector, the detector's output will go High. Some hysteresis is provided to prevent instability, should the input voltage be equal to the reference voltage. *Hysteresis* denotes the difference in response to an increasing signal versus the response to a decreasing signal. The output of gate A is normally High (gate G output normally Low). The resultant Low from gate B sets the RS flip-flop, thereby placing a High on the output of gate C. Coincident with this, two things happen: (1) QA lights the Logic Level Indicator Lamp. (2) The High outputs from gate C and One Delay cause the output of gate E to go low. This disables gate F and prevents the indicator lamp from turning off, should the input signal drop to zero during this pulse-stretching time. This cross-coupled disabling characteristic is responsible for the 10-Hz flash rate when the input is a high-frequency pulse train.

Once the High from gate C propagates through the One Delay (approximately 50 msec), a Low on the R output of gate D attempts to reset the flip-flop. The flip-flop will reset only if the input voltage drops below the logic 1 threshold level. The indicator lamp turns on to full brilliance when gate C output goes High. This occurs when the probe's input level reaches the "1" threshold. When the probe's voltage is below the "0" threshold, the RS flip-flop in the Zero Channel is set High and prevents gate J from lighting the indicator lamp. The High from the flip-flop and the High from the Zero Delay (for about 50 msec) prevent the One flip-flop from setting. When the probe voltage is not positive enough to activate the One channel or low enough to activate the Zero channel, both flip-flop outputs are Low. Both Low outputs are connected to gate J, which turns on QB and the lamp. Diode CRC provides an added diode voltage drop in the emitter circuit and causes the lamp to glow at only half brightness.

If the probe input is continuously High, the ONE channel flip-flop Out point and the Logic Lamp will be High (the lamp will be on) continuously. A single fast positive pulse, 5–10 nsec or slower, will actuate the ONE channel flip-flop and produce about a 50 msec bright Logic Lamp flash. The delay circuit stretches the pulse. A positive pulsating signal with a frequency even slightly greater than 50 MHz will cause the lamp to flash on for about 50 msec, 10 times per second. The delay circuit stretches the pulses. Negative levels or pulses will cause

the lamp to switch off for about 50 msec. If two or more termina\
logic device need to be pulsed simultaneously for test, a branched
put lead from the pulser can be employed for this purpose. The curi
capability of the pulser is sufficient to drive several input termina
simultaneously.

3.3 Troubleshooting With the Logic Clip

A logic clip, such as the one in Figure 3-15, is a very useful digital test instrument. This unit clips onto TTL or DTL IC's and instantly displays the logic states of all 14 or 16 pins. Each of the clip's light-emitting diodes independently follows level changes at its associated pin; a lighted diode corresponds to a logic-high state. The clip contains its own gating logic for locating the ground and the +5 volt V_{CC} pins. As noted previously, a logic probe is much easier to use than an oscilloscope or a voltmeter when the question is whether a terminal is in the logic-high or logic-low (1 or 0) state. Pulser-probe and pulser-clip *combinations* enable the troubleshooter to quickly answer such questions as "Is a gate functioning?", "Is a pin shorted to ground or to V_{CC}?", and "Is a counter counting?" without unsoldering pins or cutting printed-circuit conductors. A logic clip, in effect, is equivalent to 16 binary voltmeters.

Timing relationships become particularly apparent when clock rates can be slowed to approximately 1 pulse per second. Malfunctioning of gates, flip-flops, counters, and adders then becomes readily visible as all of the inputs and outputs are seen in perspective. When

FIGURE 3-15 Logic clip used to troubleshoot digital circuitry. (*Courtesy, Hewlett-Packard*)

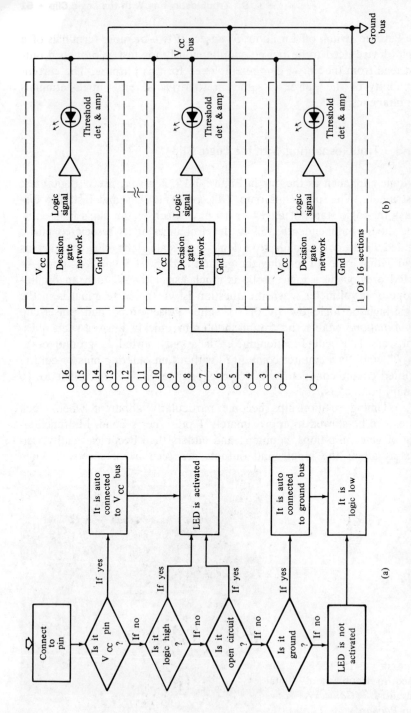

FIGURE 3-16 Block diagrams of the logic clip. (a) Decision sequence. (b) Network plan. (*Courtesy, Hewlett-Packard*)

3.4 Troubleshooting With the Logic Comparator · 63

pulses are involved, a logic probe with its pulse-stretching capability complements a logic clip—timing pulses can be observed on the probe while the associated state changes are observed on the clip. Like the logic probe and the logic pulser, the logic clip contains internal logic circuitry. Block diagrams of the logic network in the clip are shown in Figure 3-16. A logic clip cannot indicate a malfunctioning component —it can only display input and output states. Thus, the troubleshooter must understand the digital equipment under test and evaluate the test data accordingly.

3.4 Troubleshooting With the Logic Comparator

Another type of digital troubleshooting instrument, the logic comparator, is illustrated in Figure 3-17. It is designed to clip onto a pow-

FIGURE 3-17 Logic comparator. (*Courtesy, Hewlett-Packard*)

ered TTL or DTL IC and to display any logic state difference between the IC under test and a reference IC. Logic differences are identified to the specific pin(s) on 14- or 16-pin dual in-line packages with a display of 16 LED's in the comparator. A lighted diode corresponds to a logic difference. All comparison procedure is automatic. In application, a reference board with a good IC of the same type is inserted into the comparator. The comparator is clipped onto the IC to be tested, and an immediate indication is provided concerning any malfunction. Even very brief dynamic errors are detected, stretched, and displayed.

A block diagram and operating waveforms for the comparator are shown in Figure 3-18. The comparator operates by connecting the test and reference IC inputs in parallel. Thus, the reference IC is driven by the same signals that are the inputs to the test IC. In turn, the outputs from the two IC's are compared. Any difference in outputs greater than 200 nsec in duration signals a failure. A failure on an input pin, such as an internal short circuit, will appear as a failure on the IC driving the failed IC. Thus, a failure indication pinpoints the malfunctioning pin. All operating power is obtained from the test circuit's power sup-

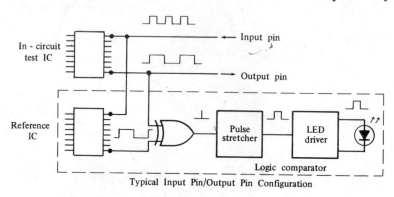

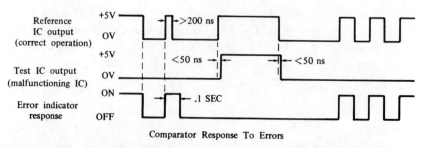

FIGURE 3-18 Block diagram and operating waveforms for the comparator. (*Courtesy, Hewlett-Packard*)

ply pins. Figure 3-19 shows typical reference boards, with one board inserted into the comparator. An advantage of a logic comparator is that it can be effectively used by inexperienced personnel or by an experienced troubleshooter who does not have knowledge of a particular digital network's circuit action.

FIGURE 3-19 Logic comparator and circuit boards. (*Courtesy, Hewlett-Packard*)

3.5 Digital Troubleshooting Techniques

Logic probes for troubleshooting TTL and DTL circuitry were discussed previously. However, high-level logic such as HTL, HiNIL, MOS, discrete, and relay logic requires a probe with logic thresholds of 2.5 volts and 9.5 volts. HTL denotes high-threshold logic, HiNIL denotes high noise-immunity logic, and discrete logic employs individual components. Relay logic is electromechanical in design and does not use semiconductors. Three types of logic probes that are required for troubleshooting these various logic families, including ECL, are illustrated in Figure 3-20. A high-level logic probe operates in the same basic manner as a conventional logic probe except that it is connected to a power supply in the 12–25-volt range. It responds to pulse repetition rates up to 5 MHz.

66 · Basic Digital Test Instruments

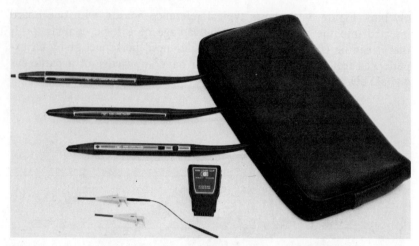

FIGURE 3-20 Three types of logic probes are required for TTL, DTL, HTL, ECL, and MOS troubleshooting. (*Courtesy, Hewlett-Packard*)

Consider the ECL logic probe. Emitter-coupled logic requires a probe with logic thresholds of -1.1 volts and -1.5 volts. The ECL probe contains high-speed circuitry that stretches single-shot pulses as narrow as 5 nsec so that they are clearly displayed. Note that the majority of IC failures involve a "dead-node" trouble symptom. (A node is a junction point or branch point). The ECL probe is powered from a -5.2-volt power supply; in most cases, it is powered directly from the circuit under test. Operation of the ECL probe is basically the same as for a conventional logic probe. Figure 3-21 depicts the probe response to various input signals. As noted previously, logic-probe tests are most useful for preliminary troubleshooting procedures. After the preliminary evaluation of circuit response has been made, it is often desirable or necessary to follow up with oscilloscope tests.

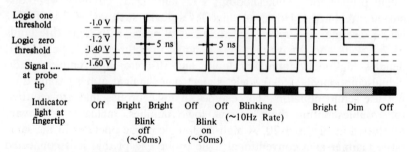

FIGURE 3-21 Response of an ECL probe to various input signals.

There are distinctive approaches involved in troubleshooting IC networks and discrete networks. When testing discrete components, simple characteristics such as resistance, capacitance, or turn-on voltages of components are verified with two or perhaps three nodes. Although the function of the system may be very complex, each component in the network performs a comparatively simple function, and normal operation is easily verified. As an illustration, with reference to Figure 3-22, each diode, resistor, capacitor, and transistor can be checked with the aid of a signal generator, voltmeter, ohmmeter, diode checker, or oscilloscope. On the other hand, when this circuit is manufactured in IC form, the individual components are no longer accessible. In turn, troubleshooting procedures involve testing the operation of the complete circuit function.

It follows that an important distinction between discrete circuitry and integrated circuitry is in the complexity of the functions (input–output relations) provided by digital IC's. Whereas resistors, capacitors, diodes and transistors must be interconnected to provide a circuit function, the interconnections of a digital IC are almost entirely internal, and the IC performs complete and complex functions. Accordingly, instead of testing and observing simple characteristics, integrated circuitry requires the observation of complex digital signals and their evaluation in order to determine if these signals are correct in accordance with the function that the IC is specified to perform.

Verification of IC operation evidently requires energizing and observing numerous inputs (there are 10 inputs in the example of Figure 3-22) while simultaneously observing several outputs (typically 2 or 3 and sometimes up to 8). Accordingly, another basic difference between discrete circuitry and digital integrated circuitry is the multiplicity of inputs and outputs that need to be energized and observed simultaneously. Over and above the need for simultaneity of signals and the complexity of functions at the device level, digital IC's introduce a new order of complexity at the circuit level. If these complex circuits are carefully studied for some time, their operation can be understood by the technician. However, this procedure is often impractical for the burdened troubleshooter—without understanding a circuit's inner characteristics, the troubleshooter needs a technique for quickly testing each device instead of attempting to isolate a fault to a particular circuit segment by tests for specific signals.

These troubleshooting difficulties can be eased or solved by taking advantage of the digital nature of the signals that are employed. Figure 3-23 exemplifies a TTL signal. The signal waveform displays voltage versus time, but in this situation, absolute voltage values are

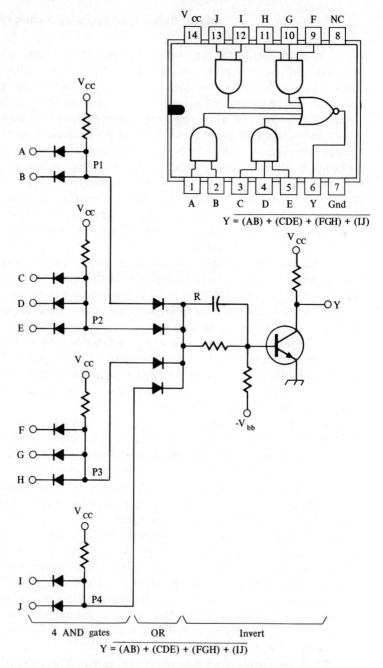

FIGURE 3-22 Discrete components are tested in terms of E, R, and C; digital IC's require the verification of complex digital waveforms and truth tables. (*Courtesy, Hewlett-Packard*)

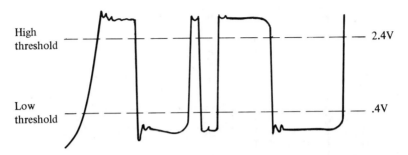

FIGURE 3-23 A typical TTL signal, with voltage thresholds indicated. (*Courtesy, Hewlett-Packard*)

unimportant. A digital signal exists in one of two or three states: high, low, and undefined (in-between). Each of these states is relative to a threshold voltage, and it is the relative value of the signal voltage with respect to the thresholds that determines the state of the digital signal. In turn, this digital state determines the operation of an IC—it is not any absolute level that determines IC operation. Referring to Figure 3-23, if the signal level is greater than 2.4, the signal is in a high state —whether the level is 2.8 or 3.0 volts is unimportant. On the other hand, in a low state, the voltage level must be less than 0.4 volt. It is unimportant what the absolute level may be, as long as it is less than this threshold voltage. In turn, when viewing digital signals with an oscilloscope, the technician must always check threshold levels.

Most digital IC's fail catastrophically. This fact makes it easier to track down a malfunction than if marginal faults were common. An important characteristic to keep in mind when analyzing a totem-pole configuration (see Figure 3-24) is that the circuit has a low impedance in either its logic-high or logic-low state. A value of 5–10 ohms is typical. Therefore, a signal source that is used to inject a test pulse at point C must have adequate power to override the low value of output impedance. A logic pulser has adequate power for this circuit test. In case the signal source is inadequate, the troubleshooter must open the circuit temporarily. For example, a PC conductor might be cut with a razor. After the test is completed, the cut is repaired with a small drop of solder.

Failures in IC networks can be classified as either internal or external to an IC. Internal failures include an open bond on either an input or an output terminal, a short circuit between an input or output terminal and V_{CC} or ground, a short circuit between two terminals

70 · Basic Digital Test Instruments

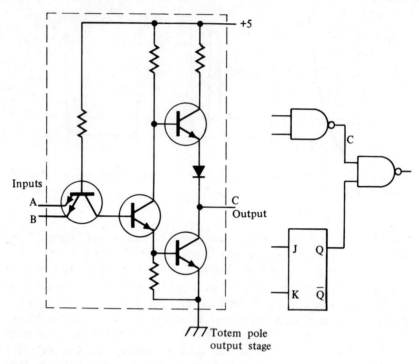

FIGURE 3-24 Terminal C in the totem-pole output stage has very low impedance. (*Courtesy, Hewlett-Packard*)

(neither of which are V_{CC} or ground points), or a failure in the internal circuitry (also termed the *steering circuitry*) of the IC. External failures include a short circuit between a node and V_{CC} or ground, a short circuit between two nodes (neither of which are V_{CC} or ground points), an open signal path, or a failure of an analog component (nondigital component). Although there are occasional exceptions, these types of faults are most common, and the digital troubleshooter should keep these points in mind when evaluating digital trouble symptoms.

3.6 Digital Symptom Trouble Analysis

It is helpful to note the effect that each of the aforementioned classifications of failure has on circuit performance. Consider an internal IC failure consisting of an open bond on either an input or an output path. This type of failure will have a different effect, depending upon whether it is an open output bond or an open input bond. In the case of an open output bond, exemplified in Figure 3-25, the inputs driven by that output are left to float. In TTL and DTL circuitry, a floating input rises to approximately 1.4–1.5 volts and usually has the same effect on circuit operation as a high logic level. Thus, an open output bond will cause all inputs driven by that output to float to a bad level, since 1.5 volts is less than the high threshold level of 2.0 volts and is greater than the low threshold level of 0.4 volt. In TTL and DTL circuitry, a floating input will be interpreted as a high level. Thus, the effect will be that these inputs will respond to this bad level as though it were a static high level.

In the case of an open input bond (see Figure 3-26), we will find that the open circuit blocks the signal driving the input from entering the IC chip. Thus, the input on the chip is allowed to float and will

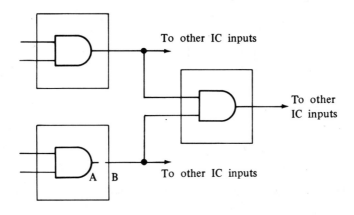

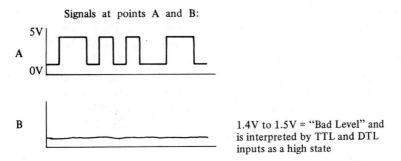

FIGURE 3-25 The effect of an open output bond upon circuit operation. (*Courtesy, Hewlett-Packard*)

72 · Basic Digital Test Instruments

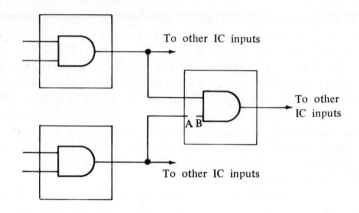

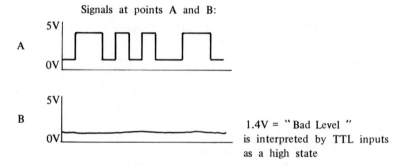

FIGURE 3-26 The effect of an open input bond upon circuit operation. (*Courtesy, Hewlett-Packard*)

respond as if it were a static high signal. It is important to realize that since the open occurs on the input inside of the IC, the digital signal driving this input will be unaffected by the open and will be detectable when looking at the input pin (such as at point A in Figure 3-26). The effect will be to block this signal inside of the IC; the resulting IC operation will be as though the input were a static high. A short circuit between an input or output and V_{CC} or ground has the effect of holding all signal lines connected to that input or output either high (in the case of a short to V_{CC}) or low (if short-circuited to ground) as depicted in Figure 3-27.

A short circuit between two pins is not as straightforward to ana-

3.6 Digital Symptom Trouble Analysis · 73

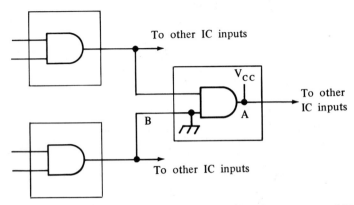

FIGURE 3-27 The effect of a short between an input or output and V_{CC} or Ground. (*Courtesy, Hewlett-Packard*)

lyze as the short to V_{CC} or ground. When two pins are short-circuited, the outputs driving those pins oppose each other—one attempts to pull the pins high, while the other attempts to pull them low (Figure 3-28). In this situation, the output attempting to go high will supply current through the *upper* saturated transistor of its totem-pole output stage, while the output attempting to go low will sink this current through the saturated *lower* transistor of its totem-pole output stage. The net effect is that the short circuit will be pulled to a low state by the saturated transistor to ground. Whenever both outputs attempt to go high simultaneously or to go low simultaneously, the short-circuited pins will respond properly. But whenever one output attempts to go low, the short circuit will be constrained to be low.

The fourth failure internal to an IC is a failure of the internal (steering) circuitry of the IC, as exemplified in Figure 3-29. This has the effect of permanently turning on either the upper transistor of the output totem pole, thus locking the output in the high state, or turning on the lower transistor of the totem pole, thus locking the output in the low state. Thus, this failure blocks the signal flow and has a catastrophic effect upon circuit operation. A short circuit between a node and V_{CC} or ground external to the IC is indistinguishable from a short circuit internal to the IC. Both will cause the signal lines connected to the node to be either always high (for short circuits to V_{CC}) or always low (for short circuits to ground). When this type of failure is encountered, only a very close physical examination will reveal if the failure is external to the IC.

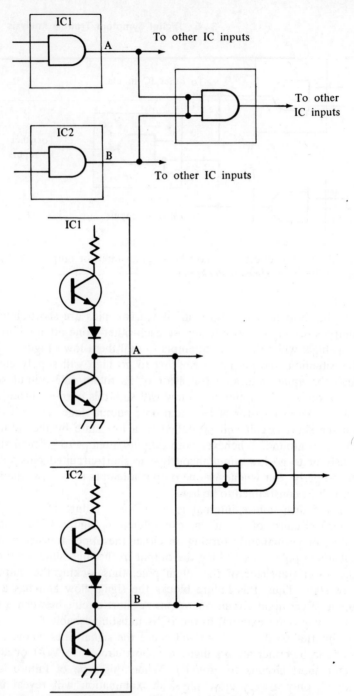

FIGURE 3-28 The error effect of a short between two pins depends on the levels. (*Courtesy, Hewlett-Packard*)

3.6 Digital Symptom Trouble Analysis · 75

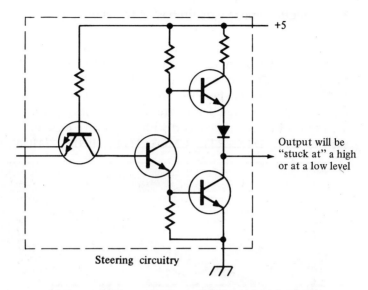

FIGURE 3-29 The effect of a failure of the internal circuitry of the IC upon circuit operation. (*Courtesy, Hewlett-Packard*)

An open signal path in the circuit has a similar effect as an open output bond driving the node (Figure 3-30). All inputs to the right of the open will be allowed to float to a bad level and will appear as a static high level in circuit operation. Those inputs to the left of the open will be unaffected by the open and will respond as expected.

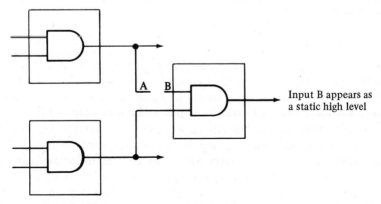

FIGURE 3-30 The effect of an open in the circuit external to an IC. (*Courtesy, Hewlett-Packard*)

76 · Basic Digital Test Instruments

3.7 Multi-Family Logic Probes and Pulsers

Logic circuit testing can be simplified and speeded up by utilizing a multi-family logic probe, shown in Figure 3-31. It indicates digital states and pulses in both high-level (CMOS) logic and low-level (TTL) logic. Logic families are selected with a slide switch. The CMOS logic threshold levels are variable and are set automatically. This type of probe can be used with all positive logic up to +18 volts, such as TTL, DTL, RTL, CMOS, HTL, HiNIL, NMOS, and MOS.

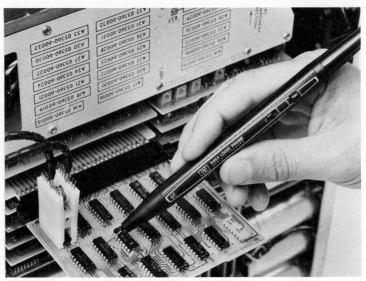

FIGURE 3-31 A multi-family digital logic probe. (*Courtesy, Hewlett-Packard*)

A current tracer, as illustrated in Figure 3-32, is an ultrasensitive tester that locates low-impedance faults by tracing the flow of current pulses rather than voltage changes in circuit conductors. By following the conducting path with the tip of this probe and watching the built-in indicator lamp, the technician can determine in-circuit logic current activity. With the ability to pinpoint one faulty point on a node, even on multilayer boards, a current tracer can be used to locate faults such as solder bridges, shorted conductors in cables, shorts in voltage distribution networks, shorted integrated circuit inputs and outputs, stuck wired-OR circuits, and stuck data busses. It senses logic current pulses as small as 1 mA, up to 5 mA, from the conductor.

Also shown in Figure 3-32 is a multi-family logic pulser with se-

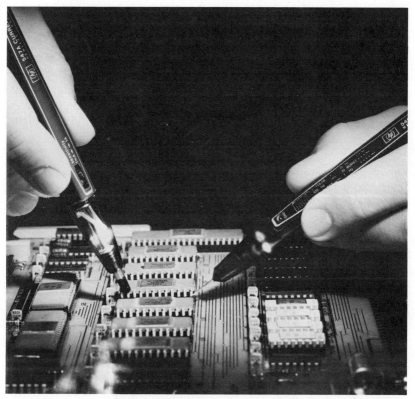

FIGURE 3-32 A multi-family current tracer and a multi-family logic pulser.

lectable output patterns. It can be used with logic families such as TTL, DTL, RTL, HTL, and CMOS. It is employed to check gates, counters, shift registers, and flip-flops. Six output pattern choices include a single pulse, a 100-Hz continuous pulse stream, 100 pulse bursts, a 10-Hz continuous pulse stream, 10 pulse bursts, or a 1-Hz continuous pulse stream. The pulser automatically drives high nodes low and drives low nodes high.

3.8 Oscilloscopes

Oscilloscopes are also basic digital troubleshooting instruments. An oscilloscope should have ample bandwidth to display the fastest and narrowest pulses in the digital system under test. Triggered sweep is essential, with a calibrated time base. A typical triggered-sweep oscilloscope used in digital-equipment troubleshooting is illustrated in Figure 3-33. An example of an input pulse that has been substantially distorted and delayed is shown in Figure 3-34. The two traces have been superimposed for comparison purposes; in practice, one pulse would be displayed below the other with a dual-trace oscilloscope. When a nonrecurrent (or seldom recurrent) pulse is to be analyzed, a storage-type oscilloscope (see Figure 3-35) is essential. Narrow noise pulses or other short-duration "glitches" often cause faulty operation of digital circuits.

"Glitches" may be caused by radiated noise, power-line coupled noise, sliver pulses owing to timing-skew input signals to gates, or other undesired circuit actions. Finding these fleeting transients can be a

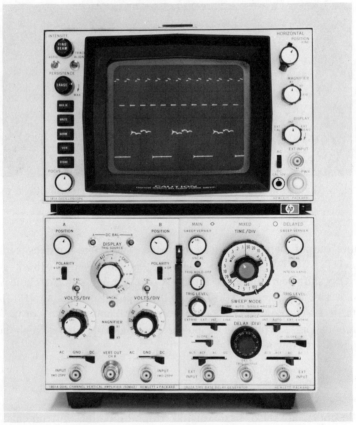

FIGURE 3-33 A triggered-sweep oscilloscope used in digital-equipment troubleshooting. (*Courtesy, Hewlett-Packard*)

3.8 Oscilloscopes · 79

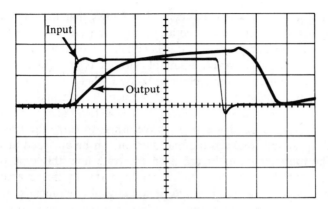

FIGURE 3-34   A substantially distorted output pulse.

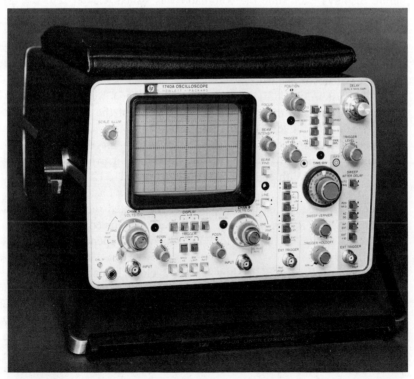

FIGURE 3-35   A storage-type triggered-sweep oscilloscope. (*Courtesy, Hewlett-Packard*)

frustrating job, even when the technician knows that they are present. When pulses occur at a low repetition rate, even most fast writing-rate oscilloscopes do not provide a display that is viewable under normal lighting conditions. However, a suitable storage-type oscilloscope easily displays such pulses. If the technician knows where to trigger and how to operate the oscilloscope for fastest storage, he can find and capture these pulses in a matter of seconds. A typical logic analyzer can store up to 16 channels of digital pulse waveforms. In turn, the waveforms that led up to and followed the malfunction can be analyzed at leisure.

The four photos in Figure 3-36 illustrate how this type of logic analyzer is used to isolate a circuit malfunction. In this example, the equipment was in the design stage. It was digital equipment that would not respond properly to a very simple program requiring it to sequence through the first 16 addresses, and then do a jump. The only way to get some signal data to study was to manually assert the Restart command. With an oscilloscope, one trace could be seen (but no other coherent display) each time the command was issued. Center-screen triggering was chosen so that events could be displayed well ahead of the trigger point, if necessary. To reduce screen clutter, only eight inputs were chosen to record and observe. The signals on other input lines could be displayed as well later, if desired, since a suitable trigger was available whenever needed.

The system clock signal ($\phi1$, $\phi2$, $\phi2$ Extend) were displayed on the three inputs, and the plan was to look at all of the address counter lines in groups of five. The analyzer asynchronous clock was set for 0.2 μsec intervals, because the shortest pulses in the defective equipment had a duration of 1 μsec. This provided four or five samples of the narrowest pulses. With eight traces being displayed, 512 samples would be taken of each of the eight input signals. From a basic knowledge of what counter line signals should look like, the circled section of address counter line A3 (Part a of Figure 3-36) was recognized to be in error. The circled section of the $\phi2$ clock on which A3 depends, corresponds to that interval.

Part c of Figure 3-36 shows the $\phi2$ clock signal positioned directly above A3 for a close comparison. The A3 pulse is shown to be ending prematurely, coincident with the first down-going pulse edge following initial turn-on. Since the counting function was performed by a microprocessor IC, it was simply replaced by a new one, and the circuit then operated properly. Thus, the defective component was pinpointed in a comparatively short time with the logic analyzer, whereas much more time and effort would have been required if some other type of analysis had been employed.

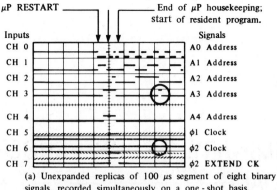
(a) Unexpanded replicas of 100 μs segment of eight binary signals, recorded simultaneously on a one-shot basis.

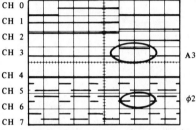
(b) Horizontal expansion of same signals as in (a) positioned to keep regions of interest on screen.

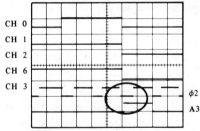
(c) Vertical expansion with φ2 positioned directly above A3. The A3 pulse should not end when it does.

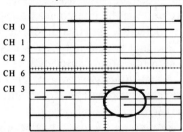
(d) Same as (c) except the defective microprocessor IC is replaced by a good one.

FIGURE 3-36 Isolation of a digital circuit malfunction by stored pulse waveforms. (*Courtesy, Tektronix, Inc.*)

4 •
PRELIMINARY TROUBLESHOOTING APPROACH

4.1 Basic Troubleshooting Procedure

In any troubleshooting process, the first step is to narrow the malfunctioning area as much as possible by examining the observable characteristics of the failure. This is often called *front-panel milking*. From the front-panel operation (or rather "misoperation"), the failure should be localized to as few circuits as possible. Then it is necessary to further narrow the failure to one suspected circuit by looking for incorrect key signals between circuits. A logic probe is generally the most useful test instrument in this procedure. In various situations, a signal will disappear completely. By probing the interconnecting signal paths, a missing signal can be readily detected. Another frequent failure is the occurrence of a signal on a line that should not have a signal. A pulse-memory unit used with a logic probe allows suspected lines to be monitored for single-shot pulses or pulse activity over extended periods of time. In turn, the occurrence of a signal will be stored and indicated on the LED readout of the pulse memory.

A comprehensive service manual for the digital equipment is important in this phase of troubleshooting. To isolate a failure to a particular circuit requires knowledge of the instrument or system and its operating characteristics. A good manual will indicate the key signals to be observed. In turn, a logic probe provides a rapid means of observing the presence or absence of these signals. After a failure has been isolated to a particular circuit, a logic probe, pulser, clip, or comparator may be utilized to observe the effect of the failure on circuit operation and to track the failure to its source (either to an IC or to a circuit external to the IC). The technique described below for applying logic test instruments such as noted above permits much faster trouble analysis than conventional general-purpose troubleshooting instruments.

Present-day digital equipment may contain several hundred IC's, and the troubleshooting problem becomes that of physically isolating the fault to the two or three IC's affected by the fault. Most of the troubleshooter's time is spent looking at signals that are good and determining that they are normal. This is not time spent in solving the problem of malfunction but, instead, is time spent in getting to the problem. A test unit such as a logic comparator provides the opportunity to drastically reduce this time and to quickly isolate the failure to only a few nodes out of the hundreds that exist in the digital system. Thus, the time that is saved can be spent to better advantage in actual analysis of the trouble.

First, with reference to Figure 4-1, a logic comparator should be used to check all of the testable IC's in the circuit, or that portion of the circuit suspected of failing, and to note the IC's and pins that the comparator indicates as failing. It is important to make certain that the comparator contacts are making a good connection to the IC terminals; otherwise, it could be falsely concluded that a normally operating terminal is open-circuited. Application of a logic comparator will rapidly focus attention on the areas of the circuit that are malfunctioning. An experienced technician requires approximately one-half a minute to check each IC.

Various IC's may not be testable with a logic comparator, or a reference unit may not be available for comparator operation. In such cases, a logic probe, pulser, and clip can be used to check IC operation. A logic probe is utilized to observe signal activity on inputs to view the resulting output signals. From this information, a decision can be made concerning the operation of the IC. For example, if a clock signal is occurring on a decade counter, and the enabling inputs (usually reset lines) are in the enabled state, then the output should be counting. A logic probe allows the clock and enabling inputs to be observed, and if pulse activity is indicated on the outputs, then the IC can be assumed to be operating normally. As stated before, it is not usually necessary to see the actual timing of the output signals, inasmuch as IC's tend to fail catastrophically. The occurrence of pulse activity is generally sufficient indication of normal operation.

Step 1

Step 2

FIGURE 4-1 Digital troubleshooting algorithm: After Step 1, the mapping step, a logic probe and pulser are used to eliminate systematically the causes of IC failures. In Step 5, the cause is determined by induction from the results of Steps 2 through 4. (*Courtesy, Hewlett-Packard*)

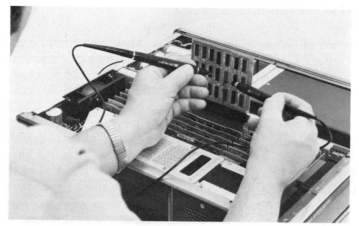

Step 3

Step 4

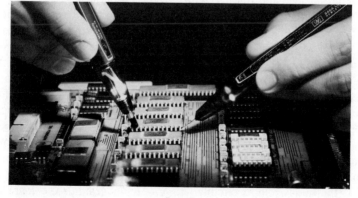

Step 5

When further tests are required, or when input signal activity is absent, a logic pulser can be used to advantage for signal injection, and a logic clip or logic probe can be utilized to monitor the response. This technique is particularly helpful when testing digital gates or other combinational devices. A logic pulser can be used to drive the inputs to a state that will produce a change in the output state. As an illustration, a three-input NAND gate, which has high, low, and low inputs, will have a high output. In turn, the two low inputs can be pulsed high with a logic pulser, and the output will then pulse low as indicated by a logic probe. This response indicates that the IC is operating correctly. A logic pulser is also valuable for replacing the clock in digital circuitry, thus permitting the circuit to be single-stepped while a logic probe and clip are employed to observe the change in the state of the circuit.

This first procedure can be characterized as a "mapping" approach, since its effect is to map out the problem areas for further analysis. It is advisable to do a complete "mapping" of the circuit before proceeding to the analysis of each of the indicated failures. Premature study of a fault can result in overlooking faults that cause multiple failures, such as short circuits between two nodes. In turn, the needless replacement of an IC may be made and much time wasted. If the technician has a complete trouble-area "map," he can begin to determine the type and cause of the failures. He does this by systematically eliminating the possible causes of digital failures, as explained previously.

A digital technician should test first for an open bond in the IC that drives the failed node. A logic probe provides a quick and accurate test for this failure. In case the output bond is open, the node will float to a bad level. If a bad level is indicated, the IC that drives the node should be replaced and retested with a logic comparator.

On the other hand, if the node is not resting at a bad level, then a test should be made for a short circuit to V_{CC} or ground. This test is easily made with a logic pulser and probe. Although a pulser is powerful enough to override even a low-impedance TTL output, it is not sufficiently powerful to produce a change in state on a V_{CC} or ground bus. Accordingly, if a logic pulser is used to inject a pulse while a logic probe is used at the same time on the same node to observe the pulse, a short circuit to V_{CC} or ground can be detected. Occurrence of a pulse indicates that the node is not short-circuited, and the absence of a pulse indicates that the node is short-circuited to V_{CC} (if it is high) or to ground (if it is low).

If the node is short-circuited to V_{CC} or to ground, there are two possible causes. First, there may be a short circuit in the network external to the IC's. Second, there may be an internal short circuit in one of the IC's connected to the node. An external short circuit should be detected by an examination of the conductors. If no external short circuit is found, the cause is equally likely to be in any one of the IC's connected to the node. Experienced technicians generally proceed to replace the IC driving the node and, if that does not solve the problem, to replace each of the other IC's individually until the short circuit is eliminated. Case histories show that a short circuit might be caused by a defective resistor or capacitor connected to the node.

On the contrary, if the node is not short-circuited to V_{cc} or to ground, and if it is not an open output bond, then the technician looks for a short circuit between two nodes. This can be accomplished in two different ways. First, a logic pulser can be employed to pulse the failing node that is being analyzed, and a logic probe can be utilized to observe each of the remaining failing nodes. If a short circuit exists between the node being analyzed and one of the other failing nodes, the pulser will cause the node being probed to change state (the probe will indicate a pulse). To make certain that a short circuit exists, the probe and pulser should be reversed and the test repeated. If a pulse is still indicated, it can be definitely concluded that a short circuit is present. A supplementary test for a short circuit between two nodes can be made by removing the circuit from the digital system and using an ohmmeter to measure the resistance between the two failing nodes. A short circuit between the nodes is easily verified in this test.

If the failure is due to a short circuit, there are two possible causes. The most likely cause is a fault in the circuit external to the IC's. This possibility can be checked by examining the conductors and correcting any solder bridges or loose-wire short circuits that may be found. Only if the two nodes that are short-circuited are in a configuration common to one IC, can the failure be internal to that IC. If, after examination of the circuit, no short circuit can be found external to the IC, then the technician proceeds to replace the IC.

On the other hand, if the failure is not a short circuit between two nodes, there are two possibilities to be investigated. The failure may be caused by an open input bond or by a failure in the internal circuitry of the IC that a logic comparator indicates has failed. In either case, the indicated IC should be replaced. An important procedure at any time when an IC has been replaced is to retest the circuit with a logic comparator. If the comparator again indicates a failure, further analysis of the problem must be made, taking into account the finding that the replaced IC is not at fault.

Another type of failure to be recognized is an open signal path in the circuit external to the IC, as depicted in Figure 4-2. This kind of failure is not indicated by a logic comparator and requires another approach. If, after using a logic comparator to test all of the IC's, none of the nodes are indicated as failing, or if after finding the cause of the failures indicated by a logic comparator, the circuit is still malfunctioning, the technician will logically suspect an open signal path.

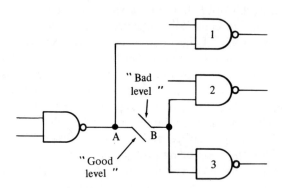

FIGURE 4-2 Effect of an open signal path external to the IC's: The open causes point B to float to a bad level, while point A is driven by proper TTL or DTL signal levels. Starting at the input of gate 3 or 4 and proceeding back toward gate 1, the exact location of the open can be determined using a logic probe. (*Courtesy, Hewlett-Packard*)

4.2 Procedural Principles

It follows from previous discussion that digital troubleshooting involves knowledge of basic logic circuits and circuit actions, availability of block diagrams and schematic diagrams, use of suitable test equipment, and knowledge of effective procedural principles. The basic essentials can be summarized as follows:

1. Know the operation of AND gates, OR gates, flip-flops, registers, and pertinent logic configurations.
2. Understand how to operate the digital equipment that is being serviced.
3. Study the block diagrams, and identify the input and output lines.
4. Determine as many symptoms of malfunction as you can in order to fully define the problem.

5. Substitute good circuit boards for suspected circuit boards whenever this procedure is possible.
6. Touch various components with your finger to find any evidence of overheating. (Do not incur shock hazards, however.)
7. Substitute good components for suspected components whenever possible.
8. Trace back along the signal path from a malfunctioning point toward the normal origin of the missing or faulty signal.
9. Test in large steps toward the origin until you find a normal signal; then test in small steps along the path of signal flow to find where the signal first becomes abnormal or blocked.
10. Try to select points for probing that are outputs of simple logic gates.
11. At a point where the signal is blocked, look for a defective component, a missing or faulty input signal, or a spurious input signal.
12. Put the original component back into its socket if it is found to be normal.
13. Make certain that boards and components are put back correctly.
14. Trace the source of missing, faulty, or spurious input signals in the same way as when starting at a primary malfunction point.
15. Probe the pins of the device—not the pins on its socket.
16. Look carefully for single pulses or low rep-rate pulses.
17. When other factors appear to be normal, check for short circuits and for excessive loads on the output of an IC where a signal is blocked.
18. If the IC is not soldered into the circuit, lift an IC output pin to isolate the load.
19. Lift IC pins to identify faulty inputs to an IC, provided that the IC is not soldered into the circuit.
20. If an intermittent condition is present, heat the circuit board to speed up the occurrence of the intermittent.
21. Cool one component at a time on a heated circuit board, and observe any change of response.

To begin isolating the trouble area in a digital system, it is advisable to test the operation of the unit. Operation is checked in terms of input–output relations. Malfunctions are analyzed and evaluated with respect to block diagrams and circuit diagrams. Although the troubleshooter might be completely unfamiliar with a particular digital system, he can acquire considerable understanding merely by noting the names of the blocks in a block diagram and by observing the names and notations for the major input and output lines. Isolation of a failure usually requires identification of the section or area that is likely to contain the fault. If the digital equipment has numerous circuit boards, the troubleshooter should first try to isolate the defective board. If replacement boards are not available, the specific defect, which is usually an IC, must be pinpointed. Sometimes a defective IC can be quickly located merely by touching a finger to it, to determine whether it is running hot. Power must be turned on, of course, and if a shock hazard exists, the test should be made with an insulated thermometer. Compare the temperature reading with that of a similar IC that is known to be operating properly.

Symptoms of malfunction should be checked under as many conditions as possible, because a clue to fault location can often be obtained by cross-checking the various input–output responses, providing a logical lead to the defective circuit board. As a practical note, there is often a tendency for the troubleshooter to skimp on the evaluation of a trouble symptom. For example, consider the common malfunction in digital equipment of missing a count. It is helpful to determine whether a certain count is always missed and whether the failure occurs in only one or in all modes of operation. Of course, no amount of preliminary symptom analysis can be expected to pinpoint the defective device or component. However, the general area of failure can be identified, after which tests must be made inside the equipment.

In the event that symptom analysis throws suspicion on a particular circuit board, this board should be replaced if a spare is available. Then, if normal operation is restored, the troubleshooter can confine his attention to pinpointing the defective device or component on that board. On the other hand, normal operation may not be restored, and the troubleshooter must change the remaining circuit boards. Boards should be changed one at a time, after replacing the preceding verified board. This procedure may not be possible owing to lack of spare circuit boards. Even when a complete set of spare boards is available, it is sometimes impossible to conclude which board is actually at fault on the basis of the symptoms. In such a case, the troubleshooter proceeds as if the equipment were one unit and makes suitable tests to track the symptom back through the circuitry step-by-step. Tracking starts at the symptom location and proceeds through pertinent branch circuits toward the origin of the missing or erroneous signal.

Tracking tests should be made in comparatively large steps whenever it is practical to do so. For example, the origin of a symptom might be a front-panel switch on the equipment. Note that it is helpful to choose test points for probing where the signal goes through fairly simple gates, if this is possible. This tracking procedure will eventually lead to a point where the signal is blocked or misdirected. When testing in large jumps, it is quite possible to skip the point where the fault is located. Once the troubleshooter finds the signal that he is looking for, however, he proceeds to follow the signal in small jumps by checking each gate in the signal path until the point of malfunction is discovered.

A signal can become blocked owing to various fault conditions—a defective component can be the cause, or another required signal might be missing at the input of an AND gate. In another common situation, signals such as preset, clear, or inhibit signals, which are normally absent at a particular time, may be present. If a plug-in component appears to be involved, it should be replaced. If the malfunction persists, or if the IC is soldered into its circuit board, the troubleshooter must employ other types of tests. All of the input and output lines including the V_{CC} and ground lines should be checked with a logic probe. If one of the inputs has an incorrect level or an incorrect signal, the troubleshooter pursues the cause of the trouble symptom as noted in the initial test procedure.

Gate identification may not be provided, and sometimes the schematic diagram or service manual does not show the internal circuitry of the IC's. In this situation, the troubleshooter must refer to the data book provided by the IC manufacturer. When probing IC terminals, it is good practice to probe the pins of the device itself—not the socket terminals—since the device pins might be making faulty contact with the socket terminals. In cases where a replacement component is unavailable, it may be possible to temporarily "borrow" a suitable replacement from another part of the equipment or from some other equipment. After a test is completed, good components should be returned to their original sockets.

Troubleshooters sometimes encounter subtle faults that demand expertise. Consider a preset or clear line which may be tripped so rapidly that the event cannot be detected with a logic probe or with a scope that is set for displaying slower signals. In such a case, the troubleshooter should double-check his findings by increasing the sweep-speed setting of the scope. Or, he may need to use the "single-sweep ready" light on the scope to obtain a definite indication of a single or low rep-rate signal occurrence. The "strobe" or "hold" mode provided on a logic probe is also helpful in this situation.

"Tricky" symptoms will be encountered occasionally in which an IC appears to be defective, although it is actually in good working condition. Several possibilities are to be considered. First, a beginning technician may be handicapped by an incomplete understanding of equipment operation. For example, insofar as symbolism is concerned, a NAND gate might actually operate as a NOR gate. This possibility simply depends upon whether positive logic or negative logic is being utilized. It must not be assumed that because one subsystem is known to operate with positive logic, that another subsystem will also be operating with positive logic. To minimize circuit complexity, designers may "switch" from positive logic to negative logic in the signal path from one subsystem to another. Note that the service data may or may not point out this type of "switch" in logic—the troubleshooter is supposed to be sufficiently knowledgeable in such situations that he will recognize at a glance the type of logic that is in use.

When the troubleshooter cannot definitely conclude whether an IC is defective or normal, he must take another approach. It is often helpful to measure the high-level and the low-level voltages. High levels in networks are never quite as high as the supply voltage in normal operation. As an example, in a TTL circuit with a +5-volt V_{CC} supply, a high level will ordinarily fall in the range from 3.5–4.5 volts, approximately. For heavy loads, the high level might be as low as 2.4 volts. Accordingly, it may be helpful to measure the voltage levels in the network. In the event that a high level is found to be the same as V_{CC}, there is probably a short circuit between the output pin and V_{CC}. By the same token, it can be helpful to measure the low levels. For instance, if the troubleshooter does not find a few hundred millivolts across an output that is in its low state, the output may be short-circuited to ground. Conversely, should a high level be measured at an IC output when a low level should exist, the output is likely to be found short-circuited to V_{CC}.

Whenever an output terminal appears to be short-circuited, it is advisable to disconnect the circuit component that drives the line, as detailed subsequently. Then, the troubleshooter can recheck for a short circuit and evaluate his finding. If normal operation is restored at the temporarily floating pin, it can be concluded that the component is normal, and the component is unplugged, its pin is bent back in place, and it is replaced in its socket. Or, the circuit-board conductor that may have been cut is repaired with a small drop of solder. As noted earlier, an output that is short-circuited may function as the input to several other components, one of which may be responsible for a short-circuit symptom. If these components are easy to disconnect, they are checked one at a time and put back correctly. It is of great importance to avoid errors in component replacement—carelessness with a circuit that is already malfunctioning will greatly complicate the troubleshooting problem. As a general rule, coping with two troubles at the same time is much more difficult than one trouble. For equipment that has been unsuccessfully serviced by another technician, it is advisable to be on the alert for double-troubles resulting from IC's that may have been plugged in backward or IC's that are wrong types.

Sometimes a short-circuited output is not caused by a faulty component or socket. For instance, solder "bridges" occasionally cause short circuits. Accordingly, the troubleshooter should carefully inspect all soldered points on the circuit board which may be too close to other conductors. It is also good practice to check any adjacent pair of solder points that look suspiciously close together. Solder "tails" or lodged pieces of wire will occasionally be discovered on close inspection. If a short circuit must be investigated further, a milliohmmeter is often helpful. If the ohmmeter reading increases along a defective circuit, the test points are progressively farther from the short-circuit point. If the readings decrease, the test points are approaching the short-circuit point. The short circuit will be found close to the point of minimum resistance reading.

A hairline short circuit is sometimes visible only under a magnifying glass. A scriber can occasionally be used to scrape away a short circuit between runs. Pin-lifting, mentioned earlier, is regularly utilized in these and other analytical procedures. (Pin-lifting denotes disconnecting an IC from its socket for test purposes. An IC is unplugged from its socket, and the desired pin is bent aside, after which the IC is replaced in its socket for test.) A pair of long-nose pliers should be used to bend the pin, and it is customary to make approximately a 45° bend. In the case of flip-flops that appear to be malfunctioning, more than one pin may need to be bent aside (if a replacement IC does not correct the malfunction). For instance, narrow pulses are occasionally fed back to the preset or clear pin by a faulty IC elsewhere, and these narrow pulses may greatly confuse the analysis. By lifting one or both of the associated IC pins, tests can generally be made that are clearly informative. Note that some lifted input may need to be tied to "high" or "low" sources. If all usual tests indicate that an IC should function although it fails to function, the output may be overloaded owing to an unsuspected fault in the load circuitry.

When the troubleshooter encounters an unusually difficult problem and feels that he is "up against a stone wall," it is advisable to wait overnight before tackling the problem again. This break in activity can be most helpful, because it allows time for the various facets of the problem and the test data to "soak in." This process is similar to the overnight insight that a student often experiences in solving a baffling mathematical problem—the "soaking-in" process allows the development of a new perspective or new angles, which often make easy work of a seemingly impossible problem.

5 •

FUNCTIONAL ANALYSIS OF A DIGITAL SYSTEM

5.1 System Evaluation

The first step in troubleshooting a digital system is to review its function. For example, the technician considers what the digital system is supposed to do—it might have a digital-voltmeter function or a clock function, or it might be part of a digital computer. Some digital systems have a self-test mode, which facilitates troubleshooting procedures. After the technician has defined the function of the digital system, he proceeds to break it down into a number of generalized or standard functional blocks. (See Figure 5-1.) These blocks will include such items as counters, registers, and memories. Memories will fall into two general categories: *read-write,* such as RAM's used for storage and retrieval, and *read-only,* such as ROM's used for microprogramming and look-up tables. Arithmetic logic tables perform arithmetical functions such as addition, subtraction, multiplication, and division. Every design of digital systems may also contain clock generators, registers, and branching.

After the digital system has been broken down into functional blocks, it is further divided into subsections. The technician proceeds to examine the combinational logic, algorithm, or microprogram for each block. An algorithm or microprogram consists of a sequence of instructions necessary to perform a given task. For example, consider the

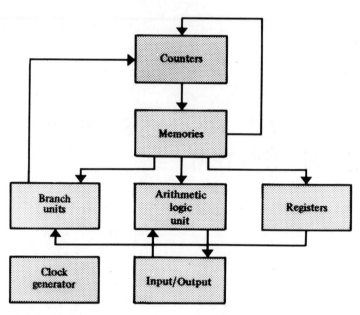

FIGURE 5-1 Function blocks of a typical computer. (*Courtesy, Hewlett-Packard*)

algorithm used to control the operations of a digital plotter. (See Figure 5-2.) This portion of the flow diagram represents the sequence used to set up the machine by setting the flag and clearing the registers. Each step is performed in sequence, and each step is numbered. The octal number assigned to each instruction is referred to as a program count or a state. A static counter allows the machine to follow the program in the correct sequence.

A ROM (read-only memory) is used to provide a look-up table or a list of instructions in a microprogram. A ROM provides a unique output for each unique input. For example, a ROM with eight outputs can provide 2^8, or 256, different patterns of 1's and 0's, as exemplified in Figure 5-3. The ROM output is usually provided by a state count, but it can come from other sources. These output electrical signals, or words, are used to activate the various blocks in a system performing the functional tasks. These blocks could include branch units, logic units, and flag or qualifier units.

The turn-on cycle in Figure 5-2 is a single case where each instruction is carried out in sequence. For more complex tasks, a break point allows the designer the flexibility of choosing alternate paths or

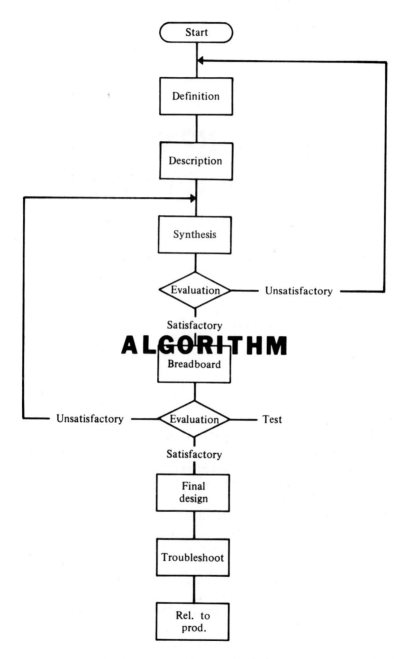

FIGURE 5-2 An algorithm is a form of flow diagram. (*Courtesy, Hewlett-Packard*)

98 · Functional Analysis of a Digital System

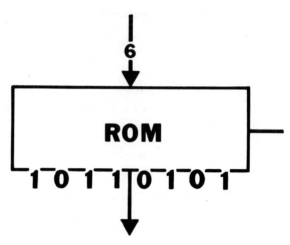

FIGURE 5-3 Example of an output of a ROM. (*Courtesy, Hewlett-Packard*)

sequences of steps for different conditions. Specific conditions are established by flag or qualifier signals. (See Figure 5-3.) These signals are recognized by a branch unit that selects which alternate instruction will follow.

Let us use the flow diagram of the plotter in Figure 5-4 as an example. At state 14, the machine checks the flag signals that indicate whether or not the button used to move the pen to the lower left corner is depressed. If it is, the ROM instructs the counter at state 15 to jump to another part of the program. If the flag indicates that the lower left button is not depressed, the machine goes on to state 16. At state 16, it either skips state 17 and goes to 20, or it jumps to state 40. Then it returns to the starting point at state 14 via the branch at state 41, thereby completing the loop depicted in Figure 5-5. The machine continues to circulate around the loop as long as neither button is pushed. A flag signal from the computer indicates that the computer is not ready to provide data.

If the branch unit determines that one of the front-panel switches is depressed, the algorithm proceeds to further instructions or states. Later, another wait loop is encountered to provide delay. Clock cycles are counted to secure sufficient delay, allowing the mechanical assembly to rise before horizontal movement begins. Similar processes are repeated through the remainder of the algorithm. The clock generator controls the timing of all operations. In the case of a plotter, the ROM

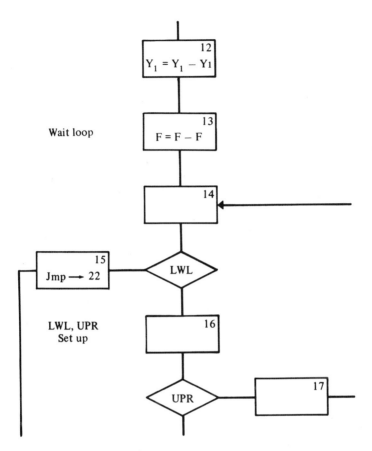

FIGURE 5-4 Flow chart at state 14. (*Courtesy, Hewlett-Packard*)

reads its inputs and provides a corresponding output every 20 clock cycles. Therefore, each instruction or step must be executed within the 20 clock cycles. The ROM strobe is 20 times slower than the clock signal. (See Figure 5-6.) At this point in the cycle, we have completed the synthesis on paper. With an evaluation of this paper synthesis, troubleshooting can begin.

We recognize that the designer chose a logic family, such as TTL, ECL, or MOS (see Figure 5-7). A trade-off was involved along with considerations of speed, cost, and ease of design. Gates, flip-flops, and other devices needed to implement each logic block of the design were selected. The ROM may have been preprogrammed from the manufac-

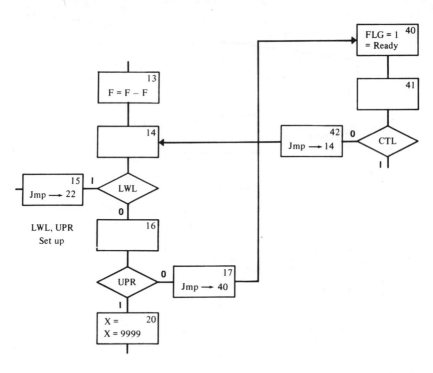

FIGURE 5-5 Completion of loop via state 41. (*Courtesy, Hewlett-Packard*)

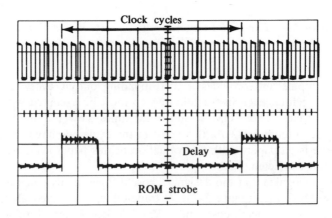

FIGURE 5-6 Twenty clock cycles are counted to develop the needed delay. (*Courtesy, Hewlett-Packard*)

FIGURE 5-7   Typical logic-family workbooks. (*Courtesy, Hewlett-Packard*)

turer, or it may have been purchased as programmable read-only memories (PROM's) and then programmed. To evaluate and debug the equipment, test and measurement instruments are first used. It's the job of test equipment to provide a meaningful picture or window of the performance of the circuits that have been selected by the designer of the equipment. The first gross indication of operation is, of course, to turn on the equipment. This is a basic check to see if some of the systems are operating. This method will work on a completed machine, but it is not very practical for a subsystem or for checking individual PC boards. If the machine is functioning, perhaps a statistical check will reveal problems. A fairly common statistical check is called *bit error rate*. It is used extensively in the data communications industry, as well as in evaluating the performance of magnetic tape or disk memories. If statistical tests indicate that a problem exists or that statistical testing is not applicable, the technician will next turn to a functional test. The technician wants to be able to see the flow of 1's and 0's. If functional problems occur, electrical analysis is necessary.

To examine functional measurements, the technician uses a logic analyzer which has been designed specifically for this function. With the device shown in Figure 5-8, the technician can examine the flow of logic at large numbers of circuit points simultaneously. Not only can

102 · **Functional Analysis of a Digital System**

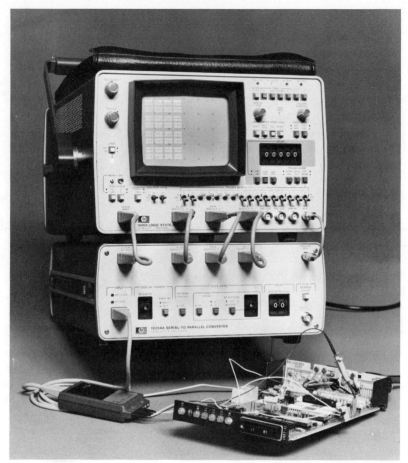

FIGURE 5-8   A logic analyzer. (*Courtesy, Hewlett-Packard*)

these parallel events be seen as they happen, but the 15 events either following or preceding the chosen trigger points can be captured. In addition to painting a functional picture, a logic analysis also enables triggering of the display on the parallel events. (See Figure 5-9.) A good example of functional testing would be to examine the flow of logic in the plotter. We connect to the ROM address so that we can monitor the microprogram and begin with the turn-on sequence. By triggering on state 1, we can see the first 11 states of turn-on and the first pass through the wait loop. Since the turn-on sequence operates properly, the wait loop is examined next. The technician may trigger

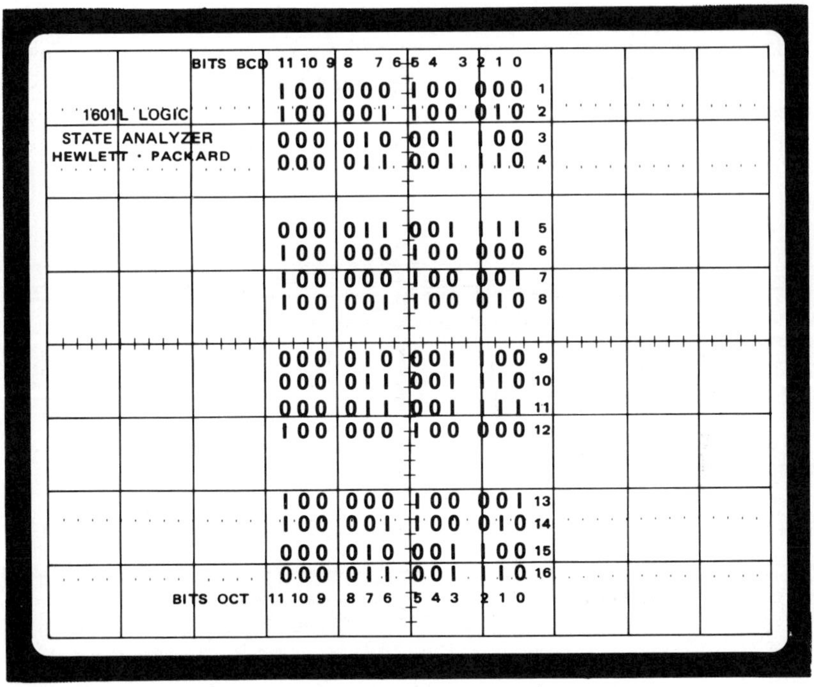

		BITS BCD	11 10 9	8 7 6 5 4	3 2 1 0			
	1601L LOGIC		1 0 0	0 0 0 1 0 0	0 0 0 1			
	STATE ANALYZER		1 0 0	0 0 1 1 0 0	0 1 0 2			
	HEWLETT · PACKARD		0 0 0	0 1 0 0 0 1	0 0 3			
			0 0 0	0 1 1 0 0 1	1 0 4			
			0 0 0	0 1 1 0 0 1	1 1 1 5			
			1 0 0	0 0 0 1 0 0	0 0 0 6			
			1 0 0	0 0 0 1 0 0	0 0 1 7			
			1 0 0	0 0 1 1 0 0	0 1 0 8			
			0 0 0	0 1 0 0 0 1	0 0 9			
			0 0 0	0 1 1 0 0 1	1 0 10			
			0 0 0	0 1 1 0 0 1	1 1 1 11			
			1 0 0	0 0 0 1 0 0	0 0 0 12			
			1 0 0	0 0 0 1 0 0	0 0 1 13			
			1 0 0	0 0 1 1 0 0	0 1 0 14			
			0 0 0	0 1 0 0 0 1	0 0 15			
			0 0 0	0 1 1 0 0 1	1 0 16			
		BITS OCT	11 10 9	8 7 6 5 4 3	2 1 0			

FIGURE 5-9 A logic analyzer displays a picture (window) of events. (*Courtesy, Hewlett-Packard*)

first on state 14. Since 16 states are displayed, the 6-state wait loop should be displayed repeatedly. If one of the front-panel buttons is pushed, the technician expects the wait loop to be broken and the algorithm to proceed to state 23. By using end-on trigger, he can look backward in time to see the 15 states preceding the trigger event.

Further states can be viewed using pattern-recognition trigger to start the display. Digital delay allows the technician to precisely index in or out 16 state windows at any number of cycles beyond the trigger point. To illustrate what happens when a problem is located by functional analysis, let us follow a typical example. Assume that when the button is pushed to move the pin to the lower left corner of the XY plotter, the pin does not move. At this point, the technician will look at the octal-coded ROM address provided by the state counter, whereupon the display shows that the machine hung up in state 15. The machine does not proceed to state 22 as it should. To isolate the electrical cause of this functional difficulty, the technician can trigger on state 15 and use this trigger point to provide an electrical analysis. This electrical analysis will usually be accomplished with an oscilloscope. (See Figure 5-10.) Functional testing can locate many faults in troubleshooting a design. It can show wrong steps in the procedure, logic hardware problems, random states that appear, and many other faults. In most cases, a functional test is followed by a detailed electrical analysis to effect a cure of the malfunction.

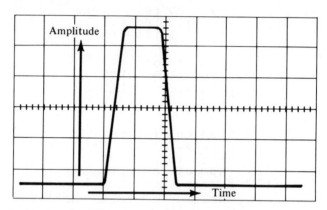

FIGURE 5-10 An oscilloscope is ideally adapted to display the amplitude-versus-time relationship. (*Courtesy, Hewlett-Packard*)

5.2 Electrical Analysis

Digital troubleshooting is concerned with timing measurements such as time-interval signal-timing relationships, propagation delays, transition times, pulse rise-and-fall times, and verification of set-up and hold times. In turn, an oscilloscope can display a picture of a condition which may be causing a malfunction. For example, Figure 5-11 shows a noise pulse; if the pulse has sufficient amplitude, it can cause a problem. An example of set-up time is illustrated in Figure 5-12. Being able to see the cause of a problem is the first step in being able to determine what corrective measures are needed. Another reason for electrical analysis is that in digital systems, this method is used to find the cause of a problem or to assist in debugging when some problem such as a jump instruction occurring at the wrong time is being analyzed. For example, the technician might find that an unauthorized pulse, or *glitch*, has caused a jump instruction to occur at the wrong time, as exemplified in Figure 5-13.

5.2 Electrical Analysis · 105

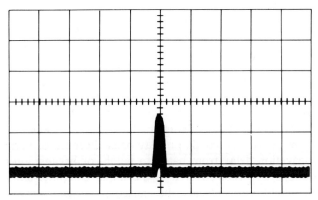

FIGURE 5-11 Display of a noise glitch on the oscilloscope screen. (*Courtesy, Hewlett-Packard*)

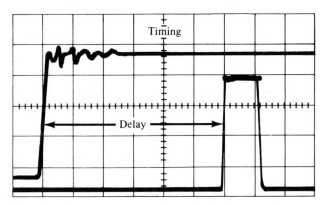

FIGURE 5-12 Example of set-up time for a device. (*Courtesy, Hewlett-Packard*)

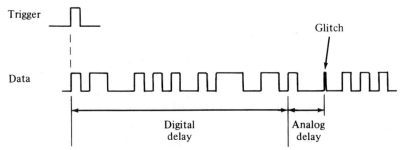

FIGURE 5-13 A glitch or noise pulse can cause early or improper triggering. (*Courtesy, Hewlett-Packard*)

A glitch will cause a malfunction if its amplitude and width are sufficiently great to cross the triggering threshold of a device. There are generally two causes for glitches: a race condition, or entry of noise. Race conditions are usually caused by timing problems at the input to a gate, as pictured in Figure 5-14. If two gates undergo transitions at the same time, or if an input changes state at the wrong time, a glitch will appear at the output. If the glitch amplitude is sufficient, another device will see it as a normal change of state and an unauthorized condition will result. A race condition may occur because of variations in device propagation delay from batch to batch, or because propagation delay variations were not covered in the initial computer-circuit simulation. Race conditions may also be caused by system noise. A glitch caused by noise looks the same as a race glitch, but it's harder to find.

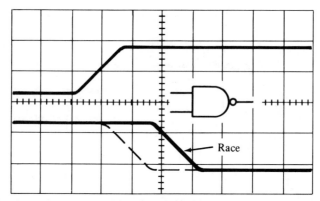

FIGURE 5-14 Race problems are usually caused by timing malfunctions. (*Courtesy, Hewlett-Packard*)

A noise glitch, as depicted in Figure 5-15, may be caused by a power-supply surge when logic circuitry changes state, by crosstalk from other data lines, or by external factors such as RFI, EMI, lights, or numerical control equipment. Finding a transient noise glitch may require a storage-type oscilloscope with a fast writing speed. Recall that functional analysis provides a convenient window that shows the flow of 1's and 0's in a digital machine. This visual flow provides an easy way to localize a malfunction. Remember too that electrical analysis provides a window (Figure 5-16) that will show why the malfunction exists. For either of these display windows to be meaningful, they must be referenced to the data flow of the system under test. In an analog system, the reference is time, in milliseconds, microseconds, or nano-

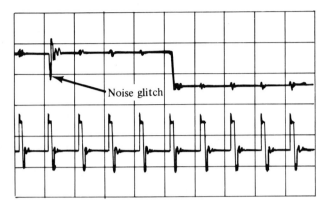

FIGURE 5-15 A noise glitch can be caused by a power-supply surge. (*Courtesy, Hewlett-Packard*)

seconds. Digital systems are referenced through events such as clock cycles, qualifiers, command signals, bit and word combinations, or specific data patterns. Therefore, the window reference will be in terms compatible with the circuit or system being tested.

There are three factors to consider when defining the display window reference in a digital system: clocking, triggering, and indexing. We must ask how the system is clocked. It might be a synchronous system, wherein everything is actuated on a particular clock-edge. Or, it might be an asynchronous system, in which events may occur independently of a clock or at two or more synchronous rates. In some cases, interface indicators, called *handshake signals,* are used to control the transfer of data between nonsynchronous units. Another factor to be considered when defining a window reference is the triggering point or the starting place. This triggering point must be unique—it must occur only once during a given system's cycle. In an analog system, a specific voltage level and slope is used as a triggering reference. In a bilevel digital system, a specific voltage level occurs over and over again, and more information is needed. A unique event must be defined. This unique event can be a specific qualifier of a command signal on a separate line or circuit node.

Since digital information is meaningful in terms of specific data patterns or unique bit and word combinations, a triggering point can be derived from the data itself. The data of interest may be in a serial format, such as PRBS sequence. Or, the data may be in parallel format, such as an eight-line interface bus. In a serial data system, the trigger-

108 · Functional Analysis of a Digital System

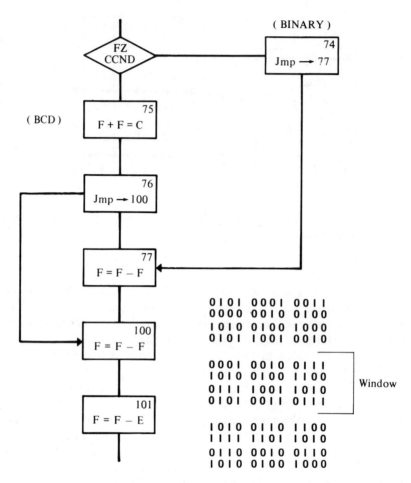

FIGURE 5-16 Example of a window display. (*Courtesy, Hewlett-Packard*)

ing event may be a specific sequence of bits, such as a preamble to a data message. To trigger a display of parallel data, the states at several different circuit nodes can be monitored simultaneously for a specific combination. For example, a functional window of parallel data is not referenced to any particular point of the input data stream. As specific bits and words are defined to form a unique pattern, the window stabilizes, and the data flow can be easily analyzed. Sometimes the data of interest are not located near the triggering point. For example, a wait list, as shown in Figure 5-17, is several cycles long, with the same

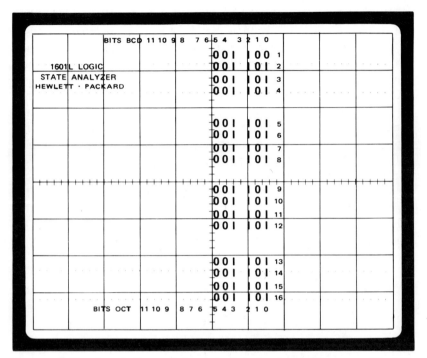

FIGURE 5-17 A wait list may be several cycles long. (*Courtesy, Hewlett-Packard*)

word occurring over and over. In order to see how the wait list ends and how long it is, the window must be indexed or moved from the triggering word to the proper place in the data stream.

As before, the index reference should be in units compatible with the system under test. In order to index the window in an analog system, the display is delayed from the triggering point by a specific amount of time. For digital systems, the delay reference is system events, such as clock cycles or word counts. In some systems, a combination of both analog and digital display may be required to properly index the display window. Other combinations may also be needed. For example, a specific serial word, such as 1111, will alert the display machine to look for a particular parallel such as 1100. Only after this combination has occurred would a display be triggered. If the data of interest occurs infrequently or only once during a data sequence, digital storage will enable convenient analysis of these events by capturing, storing, and displaying the needed data. It is apparent that a functional

110 · Functional Analysis of a Digital System

window can be used to locate a malfunction in a digital machine and that an electrical analysis window can show how the malfunction exists.

Consider the aforementioned principles in order to locate and analyze the following problem in a representative digital system. A program sequence is controlled by the outputs from a BCD counting register. Suppose that the program is not being executed properly. To verify the operation of the control register, a logic state analyzer with a functional display window is connected to the register output and triggered on the starting state of the register. (See Figure 5-18.) The display states are checked for the correct count sequence, and the analyzer display shows that these register states are correct. In order to check the subsequent states of the register, digital delay is used to index the window farther and farther into the data stream until the malfunction is localized. An incorrect count at state 89 is observed, as depicted in Figure 5-19. The functional display shows that the first decade is

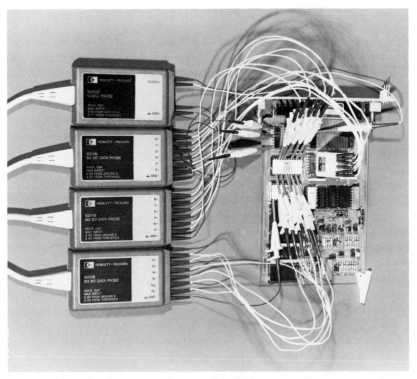

FIGURE 5-18　A connection of many data lines to an IC. (*Courtesy, Hewlett-Packard*)

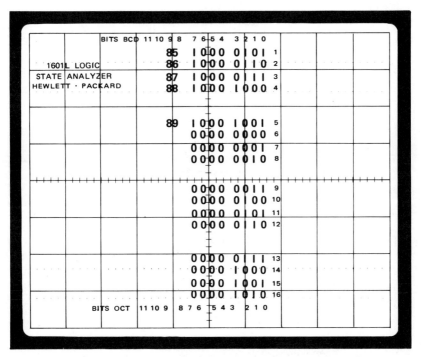

FIGURE 5-19 State 89 is identified. (*Courtesy, Hewlett-Packard*)

counting properly but that the second decade resets at state 8 and does not advance to state 9.

Inasmuch as the malfunction *is* localized with the functional window, an electrical window will be used to find the cause of the problem. Since the logic analyzer has already established the location of the fault, this reference point can be used to locate the electrical window at precisely the correct place in reference to the data stream. Since the second decade is resetting early, there could be an unauthorized pulse on the reset input to the IC. The oscilloscope is connected to the second data reset line and is triggered at state 88 with the logic analyzer. A glitch is displayed on the oscilloscope screen, revealing the reason for resetting of the second decade at the wrong time. The second decade reset signal originates at 103 of this IC, a quad two-input NAND gate. Further checking shows that the glitch is being caused by a race condition at the inputs. The signal at pin 1 reset bar is unstable and races ahead of the signal at pin 2. (See Figure 5-20.) The reset bar originates at pin 8 of this IC. A coolant applied to the IC locates the problem.

112 · Functional Analysis of a Digital System

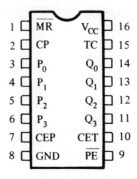

FIGURE 5-20 A fault is localized to the IC by means of a coolant spray.

Replacing the heat-sensitive IC eliminates the malfunction and returns the program sequence to normal. In this example, functional analysis was used to localize the malfunction. Functional analysis was then related to electrical analysis by triggering an oscilloscope from the logic analyzer. Finally, the IC was replaced, as depicted in Figure 5-21.

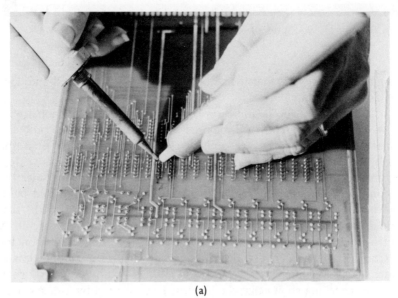

(a)

FIGURE 5-21 Replacing an IC: (a) removing solder with a solder snipper.

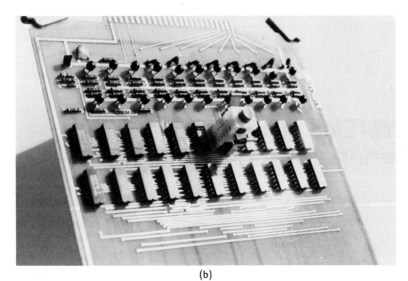

(b)

FIGURE 5-21 *Continued*   (b) IC retractor. (*Courtesy, Hewlett-Packard*)

6 •

SYSTEMS TROUBLESHOOTING PROCEDURES

6.1 Sequential Plan of Attack

Troubleshooting procedures discussed in this chapter do not involve the use of malfunction tables or symptom–cause–remedy charts. For a large interacting system, such tables and charts would be too cumbersome and inconvenient. Instead, the method suggested by Digital Equipment Corporation engineers depends upon logical thinking, common sense, and an organized step-by-step procedure. Ideally, the technician should be completely familiar with the system. When confronted with a malfunction, a technician who is not familiar with the machine wastes valuable time poring over prints and elementary system descriptions, thus unnecessarily prolonging down-time. However, in this age where the technician is expected to work on a multitude of processors and peripherals, this may amount to wishful thinking. The key here is to be familiar with the basics of the system and to know where to find more detailed information. Thus, the technician should have easy access to flow charts, functional descriptions, and the more detailed description as given in the maintenance manual. When confronting a new malfunction in a system, the following sequential plan of attack should be observed:

1. *Initial Investigation:* Gather all available information on the problem.

2. *Preliminary Check:* See if the malfunction presents any obvious physical symptoms.
3. *Console Troubleshooting:* Attempt to localize the problem to a particular section of logic; use the maintenance programs.
4. *Logic Troubleshooting:* Further localize the malfunction to a particular module, power supply, or power control unit.

In the *initial investigation,* the first step in troubleshooting a malfunction, the technician should find out as much as possible about the nature of the malfunction, before even touching the system. It is helpful to consult the log, as the machine user may have noticed some unusual machine behavior as a prelude to the existing malfunction. The same malfunction, or one related to it, may have occurred before. Perhaps it has occurred, and a record has been made of how it was remedied. The more information the technician can gather at the outset, the more rapidly he can make his diagnosis and the sooner the machine returned to operation. Every available source of information should be explored. A technician should avoid attempting to troubleshoot a computer malfunction cold; usually, this approach merely wastes time.

In a *preliminary check,* the second step, the technician should check for physical symptoms of malfunction. He should look at the console switches and verify that the user is running his program correctly. Perhaps the tape will be found faulty; a spare tape may be available. In looking over the system, he should be alert for broken cords, plugs, and tripped circuit breakers. All indicator panels should have at least some lights glowing; modules should be checked to verify that they are fully plugged in; power may be absent from a unit. This preliminary check is useful far more often in the case of a catastrophic malfunction than in the case of an intermittent problem. Except for cable and plug-in unit connections, which may be intermittent, most intermittent malfunctions are caused by cold-solder joints or by defective components. More sophisticated troubleshooting procedures are required in this situation. Nevertheless, because the preliminary check often eliminates many common sources of trouble that might otherwise be overlooked, the preliminary check should not be omitted. Few developments are more annoying than to go through complex and time-consuming troubleshooting procedures, only to discover that the malfunction is actually being caused by some cable connection that does not make proper contact.

The initial investigation in many cases discloses an appropriate line of attack but does not in itself pinpoint the location of the trouble. *Console troubleshooting,* the third step in the troubleshooting sequence, is used to localize the malfunction within a small section of the system.

The user is occasionally able to point out the failing device (i.e., "Disk doesn't work"), at which time the technician may proceed directly to that device. Sometimes the user is of no assistance (i.e., "System down"), and in this situation, the technician is forced to localize the problem on his own—probably through the use of a system-type diagnostic. Skillful use of the diagnostic can be helpful—it should be used as a tool but not depended upon excessively. It is essential to know what the diagnostic is trying to do and what is supposed to happen.

Once the malfunction has been isolated to a particular section, it can be further isolated (*logic troubleshooting*) from the console even before the technician starts to use a scope. If the technician is not sure how the diagnostic is supposed to work, he should write his own short loop: he should proceed to perform the suspected bad function, step back to observe the results, single-step the routine, and note everything that happens. In many cases, by working with the console and the prints, the technician will be able to isolate the fault to one or two gates and to find the failure by module-swapping, without any use of a scope.

6.2 Pattern Consistencies

Console troubleshooting procedures for locating catastrophic malfunctions should be directed toward discovering a pattern of consistency among the errors. A consistent pattern helps the troubleshooter to proceed logically to the site of the malfunction. For example, suppose that the preliminary investigation reveals that: "The higher-speed reader is not functioning properly." Assuming the worst case and having no low-speed reader on the teletype, the technician is forced to write a short program for the reader. This program reads one frame of tape and displays the results in the console lights. For example, putting in an alternate 1's and 0's tape results in reading 1's and halt (data lights = 377); therefore, no bits are being dropped under these circumstances. Read again: this time, the 0's are read (data lights = 0); therefore, no bits are being picked up under these circumstances. At this point, the technician concludes that it is not a case of obviously picking up or dropping of bits—he may even question whether the reader is defective at all. For example, the reader may not work at full speed, or there may be some other malfunction associated with the reader.

Next, the technician will make his program a bit more sophisticated so that the program will now read alternate 125, 252, characters at full speed. The technique consists of mixing up the data somewhat, and the reader will be run at full speed and will check the data as it proceeds. The technician toggles in a program to punch the tape and then toggles in a program to read the tape. A minor amount of debugging may precede this operation. He then runs the program. Suppose that he finds that it halts promptly with 127 in the lights—he checks his program, and he may single-step through the program. This time, the technician is supposed to read 252, but suppose that he reads 253 instead. He has now compiled the following information:

1. All 0's are read correctly.
2. All 1's are read correctly.
3. In attempting to read 125, the second LSB is picked up, making it 127.
4. In attempting to read 252, the LSB is picked up, making it 253.

At this point, the technician knows that there is a definite relationship between the LSB and the second LSB. Through further programming, suppose that he concludes that whenever the LSB is a 1, the second LSB is picked up. Next, the inputting from another device, such as a TTY keyboard, does not show the malfunction. Thus, the technician has narrowed down the problem to a very small section of machine logic. At this point, the fault could be in the reader logic from the point at which the reader buffer is gated to the input–output bus, or it could be in the logic preceding the buffer, back to and including the reader itself. This knowledge sets the stage for logic troubleshooting.

6.3 Logic Troubleshooting

After console troubleshooting procedures have localized the trouble to within a small section of the computer, logic troubleshooting procedures are employed to isolate the malfunction to within a single module or to a module connection. Logic troubleshooting is most often accomplished with an oscilloscope and small diagnostic or exercise loops consisting of only a few instructions. Logic troubleshooting is normally the fourth step in the troubleshooting sequence. It relies heavily upon successful completion of the first three steps (listed in Sec. 6–1). Logic troubleshooting is detail work, normally performed on small sections of logic or on particular discrete strings of connections between panels. With the possible exception of diagnostic and exercise loops, these procedures are applicable only to subsystems. These procedures should not be substituted for the console troubleshooting methods that have been discussed.

To avoid wasted time from widespread detail work, the technician should use console troubleshooting to isolate the malfunction to the smallest possible section of machine logic before logic troubleshooting procedures are started. The construction and development of both diagnostic and exerciser loops, as well as procedures for logic troubleshooting, are next considered. A diagnostic or exercise loop is a set of instructions, one characteristic of which causes the computer to repeat this set of instructions over and over again. A loop may contain any number of instructions and generally incorporates some method of indexing. The diagnostic and exercise loops considered here will contain only a few instructions, and they are designed to repeat indefinitely (no indexing used). Exercise loops are specifically designed to pulse some small section of the computer logic repeatedly. In general, most exercise loops contain the following three parts:

1. One or more instructions that set up the desired initial conditions.
2. The group of instructions that generates the desired pulse or level.
3. A jump instruction, returning control to the beginning of the loop.

A diagnostic loop is merely an exercise loop that includes some method of sensing for error. A diagnostic loop can be set up to halt the computer in such a way that the console indicator lights give some indication of the location of the malfunction causing the error. Good familiarity with the instruction set is required for the development of small exercise and diagnostic loops. Exercise and diagnostic loops are used generally as a way of keeping some section of computer logic operating repetitively in a predictable manner. When repetitive operation is set up in this way, oscilloscope signal-tracing techniques can be used to determine whether the correct pulses and levels are being generated. Diagnostic loops are also used during console troubleshooting procedures as an aid to further narrowing the possible trouble area.

In general, the set of instructions that sets up the desired initial conditions should be very short and certainly contain no more than three instructions. If a particular repetitive operation requires a complicated or involved pattern of initializing, then nearly all the preparation should be done at the console before depositing the exercise loop. Ingenuity, common sense, and familiarity with the instruction set, enable an alert technician to develop diagnostic or exercise loops to suit any specific troubleshooting problem. Logic troubleshooting procedures should be undertaken only after the preliminary check of console operation has isolated the malfunction to a small section of the computer logic. Logic troubleshooting is performed inside of the computer. It always consists of a number of steps designed to narrow down the location of a malfunction to within a particular plug-in unit, connection, or power unit. The specific steps required and the order in which they are carried out always depends upon the particular problem. In general, however, logic troubleshooting steps fall into three broad categories:

1. *Signal tracing.*
2. *Substitution.*
3. *Aggravation.*

Only one good method of *signal tracing* is available, aside from the logic probe, and that is the use of an oscilloscope. Since console troubleshooting has presumably isolated the trouble to within a small section of computer logic, an appropriate exercise or diagnostic loop can be used to operate the suspected section of logic repetitively. In the event of a catastrophic malfunction, the signal-tracing method determines with absolute certainty whether a pulse of good quality (amplitude, duration, and rise time) is being generated at the right time. In the case of an intermittent malfunction, this signal-tracing technique must be combined with some appropriate form of aggravation, as will be explained.

Substitution is the technique which first occurs to most technicians when faced with a logic troubleshooting task. Usually a spare plug-in module is substituted for a suspected module to determine whether the malfunction is thereby cured. When one is troubleshooting registers and counters, however, it is often more useful to exchange bits of the register or counter instead of substituting a spare module. After the exchange, if the malfunction has moved to the new location, the trouble is probably in the exchanged module. However, if the malfunction still affects the original location, the malfunction is more likely to be in some logic network that supplies pulses or levels to that location.

Aggravation, as an electronics troubleshooting or maintenance procedure, sounds as though it should be scrupulously avoided; however, it is actually quite useful. In the troubleshooting of intermittent malfunctions, aggravation is often the only technique that gives any indication of malfunction location. Some types of aggravation are vibration, marginal check (if possible), heat application, and coolant application. As long as care is taken to avoid inflicting permanent damage, the technician should not hesitate to twist, probe, worry, or poke at connections, cables, plugs, or plug-in units. Connections through plugs and sockets and through cables should be immune to any reasonable amount of pulling, twisting, or flexing. If such aggravation produces a computer error, an intermittent logic connection is probably causing the malfunction. Wiping the handle of a plastic screwdriver across the back of the suspected row of modules is often a useful technique. The resulting vibration generally interrupts an intermittent connection. By repeatedly restarting the program and narrowing the area of vibration (tapping fewer and fewer modules), the malfunction can be localized to within one or two modules. After localization of the malfunction in this way, the technician should try wiggling the suspected module up and down within the mounting panel.

Consider the application of logic troubleshooting to the aforementioned example involving the high-speed reader. To review the troubleshooting data: whenever the LSB is read, the second LSB is picked up. Whenever the second LSB is read, the LSB is picked up. The problem does not occur when the TTY keyboard is used, thus eliminating CP or I/O malfunctioning. The technician proceeds to punch a length of tape with only the LSB punched and makes a loop. He toggles in a program to read continuously and readies his oscilloscope. Looking at the reader buffer, he finds that the second LSB is set. Looking at the input to that bit, he finds the qualifying signal from the photo-amplifier. Looking at the input to the photo-amplifier, he finds the qualifying signal for the second LSB. Since this comes from the photo-diodes, he deduces that a short circuit exists between the LSB and the second LSB, and he takes steps to correct the fault.

6.4   Eight Basic Rules for Digital Troubleshooting

1. No more than two technicians should work on a digital troubleshooting task at one time. More than two workers leads to confusion.
2. Always correct obvious faults first. If there are multiple symptoms (what appears to be more than one fault), don't concentrate on the vague and unclear problem. Fix the obvious, and it may happen that the multiple symptoms disappear.
3. Always proceed with deliberate caution, and make certain that no trouble symptoms are ignored.

4. Analyze and subdivide the system. The problem may be caused by a device that is not even being used. For example, it may be helpful to pull out the cables and work with a basic system comprising only memory and the offending device.
5. Never make any unsubstantiated assumption. To assume that a signal exists, or to assume that it is a good signal *without checking* is an unjustified risk. If the technician proceeds to follow up a previous technician, it is advisable to make a complete new start from the beginning.
6. Make certain that the CP unit is working. A seemingly weird I/O device problem might be caused by a CP malfunction. Supportive CP tests should be run in this situation.
7. After several hours of troubleshooting that seemingly lead nowhere, the technician should take a break and relax for a while; he may come back with a different and successful approach.
8. After a "tough dog" has been returned to normal operation, the technician should review his work. Could it have been done faster and better? If he had it to do over again, what procedure would he employ? This self-criticism is the foundation of self-improvement and sharpening of expertise.

The flow diagrams shown in Figures 6-1 through 6-4 illustrate some approaches that can be taken in effective troubleshooting procedures. It will be recognized that some problems can be solved in a few minutes by choosing the right approach, whereas elimination of a single basic step can cause the loss of many hours of valuable time.

6.5 Development of Troubleshooting Expertise

In the view of International Business Machines engineers, no technician can become a good troubleshooter unless he works steadily to develop his expertise. A troubleshooter's mental attitude is just as important as his knowledge of the facts. In many cases, a difficult problem will appear overwhelming to the beginner who thinks of the complexity of the machine. On the other hand, with a more positive approach, his attitude toward a machine fault is: "If it's there, it can be found." With this attitude, the problem becomes manageable. It becomes apparent that a complex machine will be regarded as a group of units, and each unit will be regarded as a group of components, each of which performs a known task. In this way, the troubleshooting approach is directed away from the complex and directed toward the singular component that does not perform its normal function.

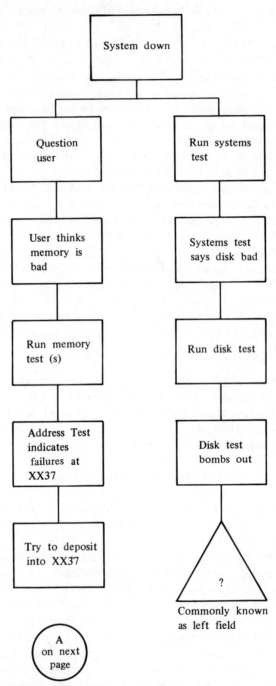

FIGURE 6-1 System-down flow diagram. (*Courtesy, Digital Equipment Corp.*)

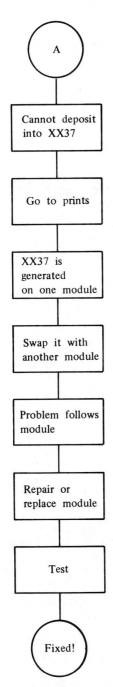

This is a fairly simple problem but you can end up in left field by omitting one of the basic steps. Following the right path, the troubleshooter logically proceeds to the source of the problem and fixes it without a scope.

FIGURE 6-2   "Cannot deposit" flow diagram. (*Courtesy, Digital Equipment Corp.*)

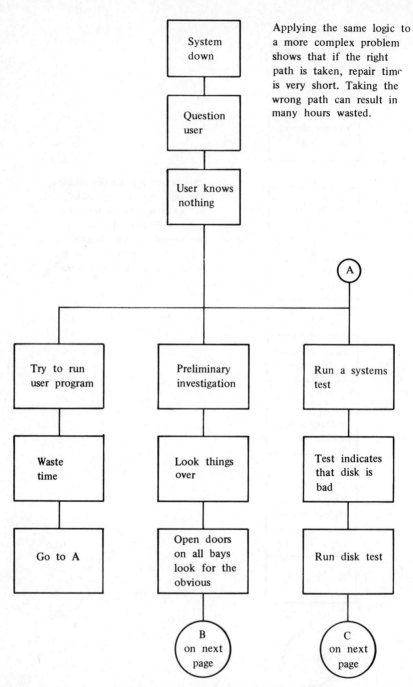

Applying the same logic to a more complex problem shows that if the right path is taken, repair time is very short. Taking the wrong path can result in many hours wasted.

FIGURE 6-3 Example of another "system-down" flow diagram. (*Courtesy, Digital Equipment Corp.*)

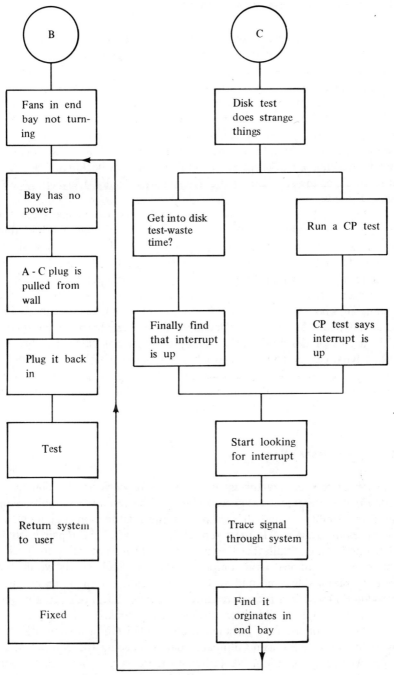

FIGURE 6-4 "System-down" flow diagram continued. (*Courtesy, Digital Electronics Corp.*)

Systematic troubleshooting is based on logical analysis. It is the opposite of what is colloquially called "easter-egging" or the "shotgun approach." Systematic troubleshooting does not exclude shortcuts, as will be explained. However, the main effort of the beginner must be directed to learning the established and prescribed test procedures and to applying them to the letter. Shortcuts develop only with troubleshooting expertise; a shortcut will often suggest itself in a particular trouble situation. Because of the essential need to know the machine and to develop an appropriate mental attitude, a beginner who tries to learn all of the shortcuts immediately may actually delay his professional progress.

Thus, a knowledge of computer principles is indispensable for the competent technician. With the advent of the solid module that performs a certain logical function, the troubleshooting emphasis is shifting from the electronic circuit to the logic-block unit. It is now possible for a technician to advance to a significant point of specialization based exclusively upon a knowledge of logic. Nevertheless, the technician who knows electronics and can read schematics can go beyond this point, because he understands the voltage levels of the inputs and outputs of the blocks. His testing procedure is safer, because he knows what can be connected to what without causing additional damage, and he can analyze whether card-swapping at a given point is legitimate or not. In addition, he can if necessary communicate with engineering personnel about engineering changes.

6.6 Job Descriptions

The job of the manufacturing technician is to perform the diagnostic tests on the equipment assigned to him. If the test reveals faults, it is his responsibility to find the reason for the malfunction and to make the necessary repairs. He must repeat the tests after the repair is completed and make certain that the machine performs according to specifications. He is charged with completing the tests and repairs within the shortest reasonable time and with communicating with his associates, particularly those in a preceding and a following shift, both verbally and by accurate and detailed written entries in a log book.

The job of the field technician, also called the customer engineer by IBM, aims at the same ultimate goal of keeping the equipment in good operating condition. Nevertheless, there is a definite distinction between the two types of activity which may be briefly expressed in this manner: The person on the manufacturing line must make a machine work which has never worked before; the person in the field must make the machine resume normal operation. A second distinction may be noted also: On the line, troubleshooting activity is essentially product-oriented—most of the man-hours of testing are spent on a single machine and operated by and with special testing machines known as robots. In the field, however, all machines are parts of a system; consequently, trouble appearing as abnormal performance of the input–output devices may originate in the central processing unit, and vice versa.

The final test technician often has arrangements very much like the actual installations in the field. In turn, he too must become less object-oriented and develop his ability to think in terms of the overall system. Apart from these distinctions, the thinking processes employed by digital technicians in locating malfunctions are practically the same as the techniques utilized to repair defects.

6.7 Symptom Analysis

The two classes of trouble encountered are: the *solid* trouble or permanent breakdown, and the on-and-off or *intermittent* trouble. Analysis of solid trouble symptoms will be considered first.

Manual test procedures are involved in production operation. Any machine coming off the assembly line must be tested according to a well-defined and documented series of steps known as the *test procedure*. A standard test procedure has its steps properly numbered according to the decimal system or to the double decimal system, which helps the user to recognize main steps and subordinate steps, as exemplified in Figure 6-5. It may very well be that after some practice, the technician may suggest ideas of his own concerning improvement of the test procedure. However, decades of shop experience have put managers on guard against arbitrary and unconfirmed deviations from approved procedures.

Some willful changes will plainly reveal the impatience of the worker. Other changes have merit, and one way to handle them is to have the worker submit a formal written proposal to be examined by the department officially in charge of and responsible for the validity of changes. Another way is to establish a suggestion plan, which may even offer monetary rewards for improvement of techniques or for removal of safety hazards. As long as the tester proceeds from step to

128 · Systems Troubleshooting Procedures

NAME:	1000 System Test	TEST ENGINEERING OPERATIONS MANUAL SAMPLE	Op Code Test		008
ITEM	8.1 Test Procedure		P/N 0000000	Page 350 - 1	

3.40   OP CODES TEST

        OPERATING PROCEDURE

.1 .0 .0   OP Codes Test

  .1 .0   PROGRAM LOAD the test deck.

        The program will WAIT and /3000 will appear in the B - Register.

  .2 .0   Turn D.E. Switch 10 ON.

        NOTE: This program is divided into two (2) basic sections.

            (a) 1800/1130 compatible OP - codes.
            (b) 1800 only OP - codes.

            With Switch 10 on each section will be run only once and the program will return to Step .1.2.0.

            With Switch 10 OFF each section is repeated 500 times and the program goes to end of test.

  .3 .0   Press START.

        The program will WAIT and /3400 will appear in the B - Register.

        NOTE: At this time the first basic section has been run once.

  .4 .0   Set Write Storage Protect Bit Switch to Yes.

  .5 .0   Press START.

        The program will WAIT and /3599 will appear in the B - Register.

  .6 .0   Set Write Storage Protect Bit Switch to NO.

A	B	C	D	E	F	G	H	I	J	K	L	M	N	O	P	Q	R	S	T	U	TOOL NO.		
																					FORM T - 198b	2 - 76	

PART 0000000

FIGURE 6-5   Sample page from a systems test procedure. (*Courtesy, IBM*)

step as instructed, he has not accomplished any troubleshooting. In other words, troubleshooting begins at the point where tests reveal abnormalities; by the same token, the official test procedure stops at this point.

When the test procedure comes to a halt, the trouble analysis begins. There is usually one symptom that signals abnormal operation; there may be other signals immediately displayed by error indicator lights, address registers, data registers, and trigger lights. Even if a very definite trouble is immediately suspected when the first faulty test result becomes apparent, the careful technician does not hastily open up the logic book and connect an oscilloscope to the suspected component. Instead, he begins what may be called a question-and-answer period between himself and the machine. He operates the keys and switches and watches the display lights of the machine. The prerequisite to what follows is, of course, a sufficient familiarity with the data flow of the machine and with the rudiments of programming the machine by feeding in information in *single-cycle* mode.

By querying the area of storage where the program stopped, the information entered by means of a single cycle will add a whole series of symptoms to the first symptom that was observed. Soon, these symptoms will fall into a recognizable *pattern*. The next step is to interpret all the symptoms and the pattern that they fall into. A pencil and paper should, at this point, prove to be important in keeping track of the results of interrogating the indicators. There is nothing wrong with suspecting the faulty component early in the test procedure. However, it would be wrong to assume that this is a sufficient basis on which to proceed with repair. Instead, the effort of the technician should be directed toward gathering more symptoms that will confirm or refute the original assumption.

It is generally recognized that the more experience a technician has with troubleshooting a variety of equipment, and the more thoroughly familiar he is with the machine in question, the greater will be the number of case histories he can solve by depressing keys and observing display panels without ever opening a logic book or connecting an oscilloscope. Quite often, an oscilloscope is used only to demonstrate once more what the technician had already known and observed as a result of his simple machine manipulations. The objection to this method—that it is fit only for the wizard who knows his machine in and out—has no basis for acceptance. If a technician does not know the machine, he is only deceiving himself; he is not going to find defects, regardless of the methods that he uses. The heart of the computer is brought directly out to the display panels on the console, and the sooner the technician knows the meaning of every one of the lamps, the more systematic will his troubleshooting methods become. If the single-cycle mode of observation must be supplemented by observations in the run mode, then it is time to employ an oscilloscope as a time-stopper and a time-magnifier. This is also the time to open up the logic book—no technician can remember the significance of the thousands of pins of a computer gate.

Next, decisions must be made concerning what pulse to use in order to synchronize the waveform under observation and what clock time to consider as a reference. Thus, gathering of trouble symptoms takes on a new dimension. To the question—What happens?—for which the technician may already know the answer, three additional questions arise: When does it happen? Where does it happen? Does it happen in the proper sequence?

6.8 Diagnostic Program

Most of the test procedures designed for line testing are not applicable to field testing. On the line, attention is centered upon one machine, though in rare cases the trouble may originate in the testing device or *robot*. In the field, the first question is: Which machine in the system is causing the trouble? Often, the customer's description of the manner in which the failure interfered with his work will assist the trouble-shooter in confining the malfunction to one of the units in the installation. If the customer can express himself only in negative generalities, it may be worthwhile for the technician to run his program up to the point of faulty performance or up to a machine error stop. The display panels will immediately display a series of symptoms.

In most instances, instead of executing the customer's program, a customer engineer will use his own diagnostic programs, often referred to in brief as *diagnostics*. These diagnostics are usually the same as utilized by the final test technician in systems testing. As a rule, diagnostics are stored on a deck of cards. Quite often, diagnostics are also recorded on paper tape or on magnetic tape. As noted previously, diagnostics are programs in which the computer and the input–output machines are required to perform all possible operations many times and to come up with error indications or message printouts in the event of abnormal operation. One of the most-often-used diagnostic decks is the *op code test deck,* the purpose of which is to exercise the op codes in the central processing unit and to determine the ability of the machine to recognize each op code and to perform the indicated operations correctly and often (one thousand times, for example) and on a rapid overlap basis.

Once the cards are placed in the hopper and the start key is depressed, the diagnostics are self-instructing; that is, messages will be printed out to tell the customer engineer what should be done next. All diagnostics will cause messages to be printed concerning completion of test and errors: some will even tell about console settings for the next test; others will have the initial setting instruction in the *diagnostics book*, which contains a detailed description of each test program and of the error indications. As an example of switch-setting, instructions for the op code diagnostic test of the IBM 1800 system is shown in Figure 6-6, wherein the positions of the switches on the main console and the customer engineering console are listed.

Switch Settings:

Main Console

Sense prog	Op mon	Dis intrpt	Chk stop	Stor prot	D/E	Mode
00	OFF	OFF	OFF	NO	0008*	RUN

CE Console

Force aux	Intrpt ctrl	CE
OFF	MAIN	00

* The D/E switch setting causes HALT on error.

FIGURE 6-6 Switch settings for op code diagnostics of the IBM 1800 system. (*Courtesy, IBM*)

6.9 Intermittent Trouble

The analysis of intermittent trouble is not essentially different from that of solid trouble analysis, because trouble symptoms do exist (even if only for a short time), and they are a valid basis for analysis. Unquestionably, intermittent trouble taxes the technician's patience, especially if he has never observed the trouble before and must rely on what others tell him. Whenever possible, intermittent trouble analysis is assigned to experienced and seasoned personnel. If the symptoms have been mildly confirmed, perhaps by fleeting observation, stronger confirmation by provoking other symptoms is difficult, because owing to the nature of the intermittent trouble, the machine operates normally most of the time. Yet, when feasible, an interpretation may be attempted by replacing the suspected part and permitting the machine to run (perhaps while other work is being done). The longer the machine performs satisfactorily with the replaced part, the greater the probability of a correct analysis.

In the field, this procedure will not be possible in most cases, and efforts will have to be made to change the intermittent trouble into a permanent trouble by introducing *marginal conditions.* The theory behind this technique is that a component is causing the trouble symptom because it is on the verge of breakdown. In turn, marginal conditions may cause the component to break down completely. Provisions for permitting the technician to introduce such marginal conditions are often built into the machine. One of these is the *marginal voltage control,* which permits (within limited extremes) an increase or decrease in the dc voltage of the power supply. Another marginal condition that may be helpful is an increase or decrease of the ambient temperature. Area by area, the temperature may be raised by cutting out fans; the temperature may be lowered by placing dry ice under intake fans (not exhaust fans).

6.10 Troubleshooting Shortcuts

A representative shortcut involves *noting recurrent faults.* Many line technicians have gone through the experience of correcting a fault in a unit on one day and then encountering the same fault the next day. Because the symptoms will be familiar to him by the second or third such occurrence, he will not have to go through the entire test procedure each time; the cause will have become obvious. One reason for faults of this kind is that if one of the assemblers misplaced a card, omitted a wire, or miswired a connection, he is likely to repeat his errors. In the field, certain malfunctions become known to field personnel in a certain model after a period of operation. The trouble symptoms are then recognized without much testing.

Card-swapping is another type of shortcut. Some technicians frown at hasty card-swapping, a term used to describe the shortcut which consists of interchanging cards. Yet card-swapping is a legitimate shortcut if done with discrimination, and with careful regard to 1) what the suspected faulty card may do in the transferred place, and 2) how the good card may be affected in the position where the trouble developed in the first place. Card-swapping is not recommended without an electronic analysis. A card may contain several circuits, of which one is not used when the card is plugged into a certain location; when put into another location, where the respective pins are used as tie points to other circuits, the introduced circuit may be damaged or may cause damage to another circuit, or to both.

Another example is a current-source card (at, say, place A) that produces a fairly constant regulated current feeding an X line driver which feeds several decode switches. A bad current driver can damage the source card at place A. Now, if a bad source at place B is swapped with the good source card at place A, it may "blow" the following driver and, depending on the address selection, several decode switch cards. If swapping must be done in this and similar cases, replacing the *entire set* of interdependent cards is indicated. However, this recommendation assumes that the tester understands their interdependence.

6.11 Learning from the Experience of Others

In field service and in manufacturing, a worker who has made a serious effort to solve a problem and feels that he is not making any progress is encouraged to ask for help. In customer engineering, IBM maintains a well-organized system of assistance. Management is so strongly convinced that a worker must learn to communicate with his fellow workers if he is going to become more competent and more efficient, that a beginner is, as a matter of company procedure, assigned to an experienced person for a period of observation and assistance with reduced responsibility. This is a time of decision for himself and for the company, in which the beginner's ability to learn from what the other person is doing is on trial. If he can profit from his fellow worker in this situation, his future requests for assistance will also be looked upon as learning opportunities and not merely as excuses for dumping his problems into another person's lap. For the same reason, the person who lends assistance must not merely busy himself to show off his skill by "fixing" the defect hurriedly, but he must give the beginner the feeling that they are both working together, searching together, and finding the fault together. Thus he is responsible for upgrading the beginner and not trying to prove the other person's ignorance.

In manufacturing operations, communication with others is a part of daily routine. Since, as a rule, a person from another shift is supposed to take over where the person of the previous shift had to stop, the problem at hand must be entered into the log book, in addition to and regardless of personal contacts between shifts. These contacts are too short; the person leaving may be tired and therefore omit important facts of his findings, and the man who takes over cannot be expected to remember such details as a series of five-place call numbers of pins or discrepancies in timing at these points when compared to a normal timing chart. The log entry must be made as descriptive and detailed as possible. Oral conversation will merely supplement this by pointing out concrete locations investigated and test equipment and tools used. It is not a rare occurrence that, having thus been informed of the problem, the next worker may find the trouble in a few minutes after the previous person had spent hours on the same "bug."

6.12 Additional Aids

In addition to the main references available in the logic books and the circuit schematics, there exists a series of troubleshooting aids of which the technician must be aware and must consult whenever they contain facts on normal functions and fault location procedures. Among these are:

1. *Manual of instruction.*
2. *Maintenance manual.*
3. *Service index.*

The *manual of instruction* is a text dealing with the normal operation of the equipment. It is written for the technician, technical trainee, and technical instructor. It is imperative reading for anyone who wants to understand the way the machine is built and the way it works. The maintenance manual contains a series of logic charts in abbreviated form, unburdened by entries concerning component locations and pin connection. These charts are referred to as ILDs (intermediate logic diagrams), as function charts, or as MLDs (maintenance logic diagrams). All of these are easier to read than the ALDs (automated logic diagrams) of the main logic book, because they present all of the important features of a particular function or operation in an illustration

6.12 Additional Aids · 135

usually confined to less than one page; as a rule, the same information is scattered over many pages of the ALDs. While a function chart (Figure 6-7) is read from top to bottom like a flow chart, the ILD (Figure 6-8) and MLD are usually read from left to right.

The *maintenance manual* also contains information on timing, either in drawn charts or in actual photographs of oscilloscope patterns,

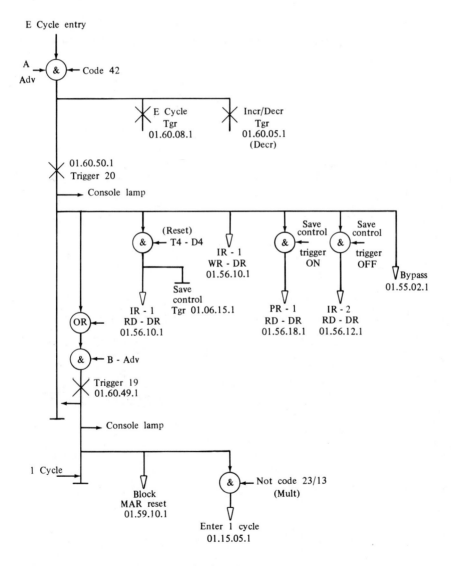

FIGURE 6-7 Example of a function chart. (*Courtesy, IBM*)

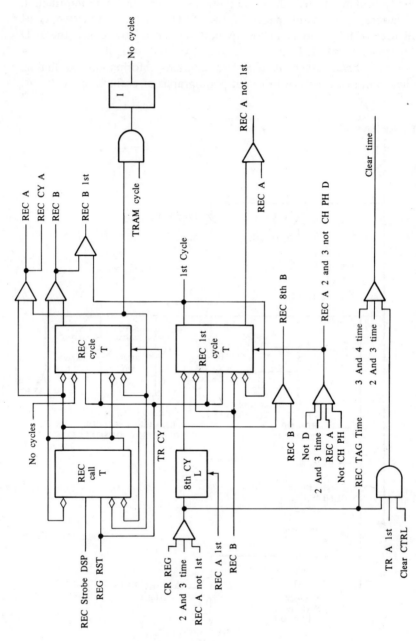

FIGURE 6-8 Example of an ILD. (*Courtesy, IBM*)

6.12 Additional Aids · 137

the latter often being supplemented with information on sync, switch setting, probe, sweep, and vertical scale. Charts for troubleshooting procedures are provided, either in a section of the maintenance manual or in a special volume known as the *service index*. These charts are read from top to bottom (Figure 6-9) like flow charts. To facilitate

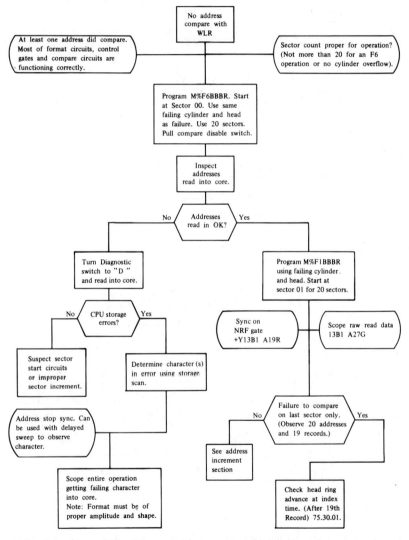

FIGURE 6-9 Wrong-length record indicator diagnostic chart sample. (*Courtesy, IBM*)

entry into the individual diagnostic charts, the service index often provides a key to diagnostic diagrams. The latter enable the technician to go as far back as depressing the start key; then, by instructions keyed to the status of the various indicator and error lights, the technician is guided into the individual chart that contains detailed steps for pinpointing the existing trouble, as exemplified in Figure 6-10.

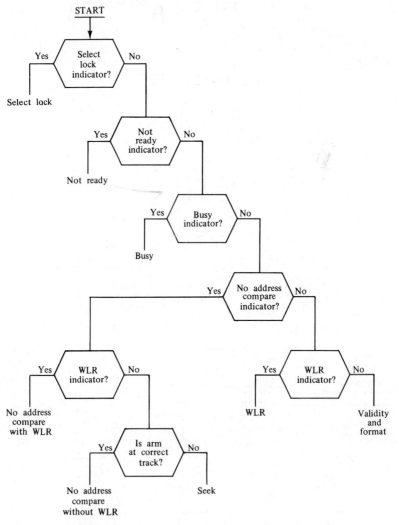

FIGURE 6-10 Key to identify individual diagnostic chart. (*Courtesy, IBM*)

7 ·
DIGITAL LOGIC ANALYZERS

7.1 General Considerations

Analysis of digital system malfunctions can be made in terms of electrical measurements or in terms of functional relations. Electrical measurements may be made with an oscilloscope in terms of instantaneous voltage values versus time. This method is helpful in analysis of noise background or noise bursts, ringing, spikes, constant dc-voltage levels, voltage swings, and so on. Most technicians have extensive experience and place greatest confidence in analysis by electrical measurements. However, complex digital systems tend to burden the technician excessively unless preliminary analysis is made in terms of functional relations. This burden results from the fact that in many cases, the traditional voltage-versus-time "window" is extremely difficult to position accurately within a digital data stream. Again, electrical viewing of signals imposes another type of burden: The signal must be interpreted to extract only the pertinent information that is required, while the technician disregards parameters that have little effect on circuit operation.

7.2 Principles of Functional Analysis

Digital information is often nonrepetitive. Moreover, extremely long (and rapid) data sequences are often encountered. Accordingly, parameters that are important in analog analysis have less significance in digital measurements. As an illustration, amplitude is an important

140 · Digital Logic Analyzers

parameter only in that the prevailing voltage must be above or below threshold values (logic HIGH or logic LOW). Also, in digital systems, time is often unimportant in an absolute sense, whereas time becomes critical when related to "system time" (clock rate of an operating system). In turn, a functional measurement consists of an observation of digital information (logic HIGH or logic LOW) versus system time (CLOCK). With this definition of functional measurement, a hierarchy of logic-state troubleshooting levels can be established. Each of these levels supplies only the information that is necessary for the particular digital troubleshooting level. Basic levels are diagrammed in Figure 7-1.

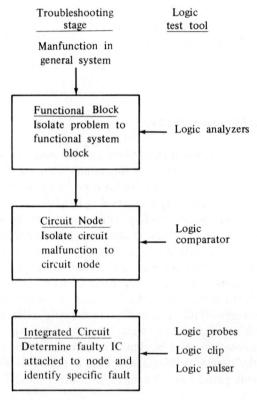

FIGURE 7-1 Three basic logic-state troubleshooting levels. (*Courtesy, Hewlett-Packard*)

7.3 Logic Analyzers

To troubleshoot digital circuits effectively, four fundamental requirements must be observed.

1. A *display* must be provided. This can be compared to a window or a picture utilized by the technician to view logic action.
2. A *reference* is required. This is simply a trigger point that is sensibly related to the data that is to be displayed, or it is a method of defining a unique trigger point within a data sequence.
3. An *indexing* method is necessary. This designates an ability to move the display window back and forth in time with respect to the reference.
4. *Storage* is essential. The test data must be stored, and it is necessary to "call back" single-shot events for detailed analysis.

From the viewpoint of conventional test methods, large numbers of "channels" (sets of data lines) are cumbersome to interpret when the troubleshooter is interested only in logic states versus system time. However, from the viewpoint of functional relations, logic analyzers such as the one shown in Figure 7-2 minimize this factor by displaying a functional picture of groups of data bits (digital words) versus the system clock. This presentation is very easy to evaluate whenever functional relationships are under examination. Displays provided by a logic state analyzer are in terms of logic HIGH's (1's) and logic LOW's (0's) versus clock cycles. Referencing is accomplished by means of

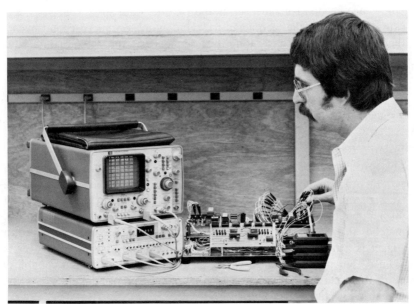

FIGURE 7-2 A logic state analyzer. (*Courtesy, Hewlett-Packard*)

trigger switches; these switching facilities permit selection of a stipulated trigger point. Furthermore, a logic state analyzer permits the display to be moved in time from the trigger point by means of a digital delay line in either a positive or a negative direction of time.

Digital delay in negative time is provided via the inherent storage facility of the logic state analyzer so that the troubleshooter is enabled to analyze a number of events leading up to a selected trigger event. A typical logic analyzer can display up to 64 bits (in serial access mode) of data that occurred before the trigger point. Digital delay in positive time permits movement of the display downstream from the trigger. As an illustration, in a disc memory, the start of a sector may be the only available unique trigger point, whereas the data to be analyzed may be thousands of bits downstream from the trigger. A logic state analyzer with digital delay can position the display window precisely, and without jitter, at the exact location of the character or signal to be examined.

In digital systems, very low rep-rate or single-shot events will be encountered that require storage to permit analysis. For instance, "once per keystroke" calculator sequences fall into this category. A logic state analyzer contains adequate memory capacity to capture and store such events; thus, the analyzer is extremely useful for analysis of single-shot circuit actions. Digital triggering and delay are necessary for functional analysis, and these features are also of great value when "aiming" or positioning electrical analysis windows on oscilloscopes. These capabilities are needed for both serial and parallel data stream analysis, because they permit the technician to "window-in" on microsecond events that occur as part of very long data sequences. And a logic state analyzer can provide trigger signals for electrical analysis of pulses, if required.

7.4 Triggering Serial Data

In serial data analysis, data pattern recognition may present a problem. This problem can be solved if the data or instruction portions of a serial word are known. In turn, it becomes possible to generate a unique trigger from the known serial event. Thus if a pattern that has been set on a digital pattern analyzer, such as the one shown in Figure 7-3, matches the bits contained in the instruction portion of a serial word, a trigger is generated. (See Figure 7-4.) Accordingly, a unique trigger is defined that permits analysis of serial data streams, and the capability of digital delay is provided, which allows further indexing from the operator-selected trigger point.

7.4 Triggering Serial Data • 143

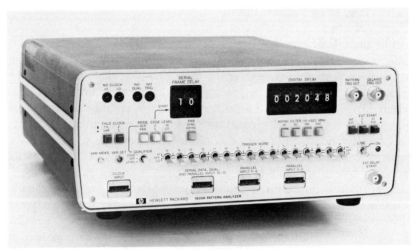

FIGURE 7-3 A digital pattern analyzer. (*Courtesy, Hewlett-Packard*)

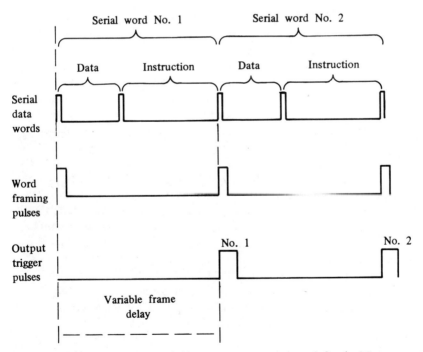

FIGURE 7-4 Frame delay and bit pattern are operator-defined. (*Courtesy, Hewlett-Packard*)

7.5 Triggering Parallel Data

In the analysis of parallel data, it is often necessary to trigger on the simultaneous occurrence of several events. As an illustration, if one or more channels of data go high at the same time that the CLOCK signal goes high, a trigger could be generated at this point in time. Also, the selected trigger events could be either high or low polarity signals. Triggering need not be clock-related and can be asynchronous if desired. This facility permits the technician to initiate the display sequence on a signal that might not be present when the clock samples the inputs to the logic state analyzer. Signals such as spikes or other random events can accordingly be captured and displayed. After a digital circuit malfunction is detected by the troubleshooter with a logic state analyzer, other digital test instruments may be utilized for more detailed tests in the system or in a circuit. After a fault has been isolated to a particular circuit area or circuit board, a logic probe, logic pulser, or logic clip can be applied to test digital IC's in the circuit.

7.6 Logic State Analyzer Characteristics

The logic state analyzer illustrated in Figure 7-2 provides 12 parallel data stream measurements at clock speeds up to 10 MHz. Data bits in 1 and 0 character form are written horizontally and correspond to the data points where the data probes are connected into the system under test. The 12 channels (8-bit words with 4 qualifiers) are displayed vertically in synchronization with 16 consecutive clocks or strobes, thereby maintaining system timing and data relationships. For easy interpretation, the display can be formatted in octal groups of three or four to match the system under test. A logic sense switch is provided to accommodate either positive or negative true logic systems.

Triggering occurs in clock synchronism when data matches the preset word with the 12 parallel trigger switches. Triggering capabilities in this type of analyzer are quite flexible, so that the instrument can access virtually any desired 16-word sequence in the data stream. For instance, the trigger word can start the display, stop the display to show what occurred before triggering, or start a counter to delay the display by any preset number of clock pulses (up to 99,999) after the trigger word. A start-display triggering mode may be selected by the troubleshooter to trigger the analyzer on a unique trigger word and to display the trigger word with the 15 following words as they are clocked through a digital system. This start-display mode is very helpful in paging

7.6 Logic State Analyzer Characteristics · 145

through a system while following an algorithm to verify operation or to quickly determine where a wrong signal path is taken. A "mark" pushbutton is provided on this instrument to increase the trigger word intensity for quick recognition of where the analyzer triggered.

As noted previously, the 16-word memory employed in this type of analyzer captures events leading up to and including the trigger word, thereby permitting the troubleshooter to inspect "negative time." This negative time capability, coupled with single sample operation, is very helpful in troubleshooting procedures, because the technician can trigger on an illegal (unallowed) state to see what led up to the malfunction. Fault isolation is thereby facilitated, because the troubleshooter can analyze the reason for error occurrence rather than merely the error itself. When the data that the technician wishes to see does not immediately follow the desired trigger word, digital delay positions the 16-word "window" an exact number of clock pulses (0–99,999) from the trigger word. Digital delay is useful for moving the display window past wait loops, monitoring the contents of tape or disc memories, and measuring lengths of subroutines.

Free-run operation in a digital analyzer is equivalent to setting all trigger switches to OFF, which clocks the display in synchronism with the clock input. This mode of operation aids in troubleshooting a system by clearly displaying active (superimposed 1's and 0's) and inactive (unchanging) data lines. When a display is not present, the No Clock and No Trigger indicators quickly pinpoint the problem and tell the technician what is preventing a display. The No Clock indicators inform the troubleshooter when the clock input does not cross the logic threshold and will indicate if the clock is above (Hi) or below (Lo) threshold. A TTL threshold selection provides the operator with convenient compatibility in testing systems that use TTL logic; also, a variable threshold selection permits adjustment over a $+10$-volt to -10-volt range. Absence of trigger signal is shown by the No Trigger indicator, which signifies that the input data do not match the Trigger Word switch setting.

Extension of troubleshooting capabilities in digital circuit analysis is provided by pattern-trigger and delayed-trigger outputs of the instrument. A pattern-trigger output signal is generated each time that the pattern set on the trigger-word switches occurs, and this signal can be used to synchronize an oscilloscope for electrical analysis at the digital point when a malfunction occurs. A delayed-trigger output signal is generated coincident with the first displayed word. In the Start Delay mode, this delayed output can be used to synchronize an oscilloscope with data that do not occur immediately after the trigger word.

7.7 Logic Analyzer Operation and Application

Next, consider the characteristics of the logic analyzer illustrated in Figure 7-5. This type of instrument provides a unique digital analysis capability by enabling the troubleshooter to "see" data at a circuit node exactly like the digital circuitry being examined, with the same timing relationships and format. A totally digital display is provided by the instrument, with amplitude being expressed in a digital format (logic highs and lows) and with time as digital time relative to defined clock transitions (clock cycles). Thus, the technician approaches a digital malfunction in its own domain (the data domain). The key to this data domain is the utilization of the time base of the digital system under examination as the time base for the analyzer. This ability to see the data in the same timing diagram format as the system under examination enables rapid functional isolation of a circuit malfunction to the basic gate or other circuit element causing the trouble.

FIGURE 7-5 Appearance of a logic analyzer. (*Courtesy, Hewlett-Packard*)

Data inputs to this logic analyzer have an input impedance of 1 megohm and accept data in the same manner as a D flip-flop or shift register. Note that the analyzer samples the input data on a defined clock transition (edge) and displays it in terms of bits referenced to the clock in the system under test. These 1's (on LED's) and 0's (off LED's) can be directly compared to the timing diagram of the circuit node, as shown in Figure 7-6. The data display of this logic analyzer consists of two rows of 32 LED indicators, with each indicator representing the logic state of the signal during a particular clock transition. In turn, the instrument provides display flexibility to the troubleshooter via two separate display channels (windows), each with a 32-bit capability. If a larger display window is needed in a demanding application, another mode may be employed which extends the A channel capacity to 64 bits.

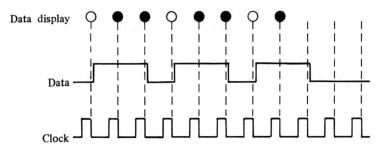

FIGURE 7-6 Example of data display with respect to data signal and clock pulses. (*Courtesy, Hewlett-Packard*)

In any display of information with respect to time or events, there must exist some unique sync or starting point for the display. Definition of this point in a digital waveform requires a deviation from the traditional negative- or positive-slope triggering technique. This logic analyzer employs a new digital triggering format, which allows indexing to any position within a data sequence by selecting a signal at either the A or B input, at the external-trigger input, or at logical AND combinations of two or three of these inputs or their complements. If more than three data inputs are required to define a unique starting point within a sequence, the parallel triggering capability of this logic analyzer may be expanded greatly by means of auxiliary probes.

Another unique mode of display position reference provided by this instrument is "asynchronous triggering." This triggering technique permits the troubleshooter to initiate a display sequence on a signal that is not present when the inputs are sampled on the selected clock transition (not accessible to synchronous triggering). If the desired data display is not present immediately following the trigger, the variable digital delay of this analyzer allows repositioning of the display to any point within the data sequence. The 32-bit "display window" can be moved with digital precision an exact number of clock pulses relative to the fixed trigger point. Data occurring far downstream in a data sequence become conveniently visible by dialing the appropriate delay number into the front-panel thumbwheel delay register.

148 · Digital Logic Analyzers

This type of logic analyzer also provides a look-ahead or "negative delay" function. Thus, the instrument always has access to the last 64 bits of data prior to the occurrence of the trigger and has the ability to display these data if desired. Accordingly, not only can a failure mode be observed, but also the sequence of events that lead to the failure can be displayed for analysis. The digital nature of this type of analyzer makes single-shot storage an inherent capability. When the troubleshooter places the instrument in its "STORE" mode, both input channels will capture the next data sequence and hold these data until reset. If selective storage is desired, the "STORE B" mode may be employed. In this mode, one channel of data may be held while the other channel continuously collects new data. This mode may be used to store a reference of known good data to which succeeding data can be compared.

A special troubleshooting capability provided by the type of logic analyzer is its ability to detect spikes as narrow as 15 nsec between clock pulses in a data stream. When operated in the "SPIKE A" mode, the instrument ignores synchronous data and only indicates the location of spikes. These spikes may be caused by race conditions, ringing, noise, or design, and are defined as more than one transition of the data on the A channel between clock cycles. This "SPIKE A" mode, used in conjunction with the digital DELAY, can be used to look for spikes anywhere in a long serial data stream, even on a single-shot basis. Figure 7-7 depicts a typical indication of spikes.

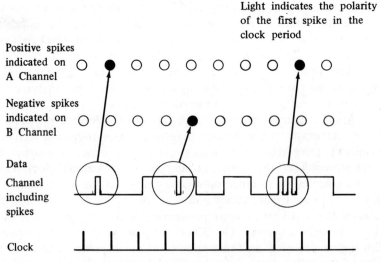

FIGURE 7-7 A typical indication of spikes by a logic analyzer. (*Courtesy, Hewlett-Packard*)

7.8 Pattern Analyzer Operation

A digital pattern analyzer was illustrated in Figure 7-3. It is instructive to consider the characteristics of this type of instrument. This pattern analyzer generates a trigger from serial or parallel digital pattern recognition and/or digital delay for oscilloscopes or other externally triggered instruments. Pattern recognition is selectable up to 16 bits in either serial or parallel mode, with digital delay selection up to 999,999 bits. A separate qualifier line is provided for use in the serial mode, which permits the troubleshooter to look for bit patterns at a discrete time or during time intervals. A serial frame delay gives the technician window selection in the bit stream relative to the qualifier starting edge.

In the parallel mode of operation, this type of analyzer is capable of either synchronous or asynchronous operation. In its parallel asynchronous mode, a selectable pulse-width filter reduces the possibility of false triggering caused by glitches resulting from skew in the data stream entering the instrument. Digital delay can be started by pattern recognition or by an external trigger input (Ext Delay Start). This permits the troubleshooter to move the measurement window a selectable number of clock cycles downstream from a uniquely selected trigger point defined by the analyzer or by the trigger input.

8.

TESTING OF DIGITAL AND PRINTED-CIRCUIT BOARDS

8.1 Functional Testing and Fault Isolation

When testing digital and printed-circuit boards in a manufacturing process or in a maintenance situation, the technician is dealing with a module or board containing printed wiring, which interconnects replaceable (via sockets or unsoldering) digital integrated circuit devices that are dual in-line, flat pack, or TO configuration. The problems involved are:

1. Testing the PC card to ensure that it operates correctly in the "system" in which it is intended to be used.
2. Finding the location of any failures, whether they are in a replaceable package or a wiring trace.

The term *functional test* is usually applied to the first problem; the term *fault isolation* is generally applied to the second problem. When working with analog circuits, a functional test is often a long sequence of measurements which attempts to exercise all functions expected to be performed. With digital circuits, this procedure is likely to be impossible or undesirable. The number of 1's and 0's needed to exercise all functions of a board can be astronomical. Thus, in a simple combinatorial pattern, 2^n patterns are required for exercising all possible input combinations, where *n* is the number of inputs. Fortunately, the number of possible failures of a digital circuit is limited. A circuit will fail in a "stuck at" condition, owing to a broken bond, a short circuit, or an open trace. Because logic circuits are designed to be unstable between states, a "stuck in the middle" condition is rare. Knowing the number of internal circuit nodes, we can account for possible failures as follows (see Figure 8-1):

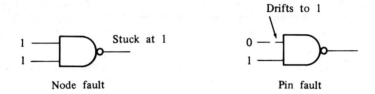

FIGURE 8-1 Possible digital circuit failures.

Node Fault: Node stuck high or low.
Pin Fault: Node "split," owing to broken track or bond.

Consider the simple combinatorial circuit shown in Figure 8-2. Note that the diagrams in this section are documented in a format that can be coded for simulation. The vertical lines represent a single connector, labeled C, which represents the circuit boundary. Connector pins are labeled A, B, C, etc.; gates are labeled G1, G2, etc.; flip-flops are labeled FF1, FF2, etc. A commercial device will be labeled U1, U2, etc. Nodes are labeled in terms of a driving device and pin. Thus, the node controlled by G1 pin 1 is G1-1. (C-C is the node controlled by connector pin C.) Pins are numbered counterclockwise beginning with the output, unless numbers on a commercial device take precedence.

With reference to Figure 8-2, there are five circuit nodes, including those on connector pins. Four of these are input pins to gates and could be broken, either owing to open traces or to bonds. We could

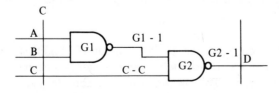

FIGURE 8-2 A simple combinatorial circuit.

consider short circuits among adjacent pins and traces owing to solder splashes; however, this is not necessary as will be developed subsequently. Counting all node and pin faults in this instance, we obtain a total of 18 faults that are either stuck-at-1 or stuck-at-0. If all of these possible faults are detected by some test, we can assert with confidence that the board is good. It is true that intermittent failures can occur as well as oscillating failures, stuck-in-the-middle failures resulting from feedback conditions, temperature-dependent effects, and other problems. Nevertheless, the "stuck at" model is, from experience, a very good approximation of reality—in fact, most of the aforementioned types of failures can still be detected, although locating them presents unusual problems.

The approach that has been adopted is to define possible faults and to systematically test for their existence. If all the tests are passed, the board is considered to be fault-free, and it is expected that, if properly designed, it will perform its intended function. Note that in a few cases this involves testing for faults that do not affect circuit operation owing to unused functions or redundancy. The circuit shown in Figure 8-2 can be tested using five patterns—all eight possible input combinations need not be applied—and an accurate measure of test quality, or efficiency, can be developed. This is the percentage of total faults that can be detected by an input pattern set. If all five patterns are applied in the case of Figure 8-2, a 100% test is provided. Or, if four of the patterns are applied, and they detect 14 of the 18 faults, there results a 14/18 × 100, or 78% test. On a board of any size, a means is needed of automatically identifying possible faults and evaluating whether they are detected. These are the principal functions of the logic simulator. A simulator is the only means by which faults can be accurately counted and, therefore, is the only means by which a high-quality test can be achieved.

Consider a printed-circuit card containing a fault. To detect the fault, two conditions must be satisfied:

1. The node in question must be set to the state *opposite* the fault being detected.
2. The change (or lack of in the event of the presence of a fault) in state must be detected.

The required approach to fulfilling these conditions is to sense every node, thereby immediately communicating internal information to the test equipment. This, in turn, entails elaborate and expensive test adaptors. Such adaptors are called "bed-of-nails" fixtures. This form of adaptor simplifies test programming, because all nodes are sensed, and it is necessary only to move each once. Testing is fast because relatively few patterns are required and probing is not needed. However, there are several disadvantages to be noted:

1. Unless a simulator is used, there is no measure of test quality.
2. One adaptor per board is required; tooling costs can be very high if a wide variety of boards is required.
3. As logic becomes more complex, less of the circuit is probed. Testing for failures in a device such as a ROM or RAM is difficult.
4. A large number of pins is required, which means that the test system can be costly.
5. Fixtures are unreliable and costly to maintain.

It is desirable to test a board from its edge connectors. This requires that in addition to moving the suspect node, a path be set up that communicates the failed node to the output. This is illustrated in Figure 8-3. Assume that the fault of interest is G2-1 SAØ (this is shorthand for "the signal driven by G2 pin 1 is stuck at zero"). If the state of node G2-1 is to be communicated to output D, gate G1 must be "sensitive," meaning that input pin 3 is a logic 1. To obtain this condition, connector pin C is set to a logic 1. If an input pattern detects a fault, the output pattern will differ from that calculated for a good circuit. With reference to Table 8-1, patterns are checked for their ability to detect the fault G2-1 SAØ.

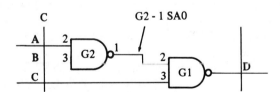

FIGURE 8-3 Communication of failed node to output.

TABLE 8-1

	FAULT PRESENT		FAULT ABSENT	
	A B C	D	A B C	D
Pattern 1	Ø Ø Ø	1	Ø Ø Ø	1
Pattern 2	Ø Ø 1	Ø	Ø Ø 1	1

Pattern 1 does not detect the fault, since output pin D is identical in both cases. Pattern 2 does detect the fault. A logic simulator models a good circuit and many bad circuits, each containing a single fault. If the outputs differ, the fault in the bad circuit has been detected. Using this technique, a program can absolutely determine whether any of the possible faults in a circuit have been detected by a particular set of patterns and can keep a statistical count of "undetected faults." The test engineer now knows whether additional patterns must be developed to bring the test to an acceptable quality. From 90% to 95% constitutes an acceptable minimum in most cases.

G1 may be contained in a different package from G2, G3, and G4. It is difficult to determine whether G1-1 or G2-2 are SAØ. Either of these faults could short-circuit the entire node to ground, and it is not possible to determine the cause without removing a device. In most situations, replacement of both packages is the advisable action.

8.2 Fault Isolation

Detecting a fault, in the present context, means that we know that there is at least one fault in the PC board. Fault isolation is the process of determining in which IC or other replaceable or repairable component the fault is located. In this detection, a "trail" of failed nodes exists from its location to the output. In a combinatorial circuit, this occurs on a single input pattern, viz., the one on which a failure has been detected. In Figure 8-4, the pattern shown detects the fault G1-1 SAØ. The nodes G2-1 and G3-1 lead from the fault location to the output at pin E. In a sequential circuit, the trail may not exist on the same pattern. If a three-bit shift register were in series with G3 pin 1 and pin E, the faulty node on pin E would show up three clock pulses after the first pattern that detects it, assuming all following patterns clocked the register.

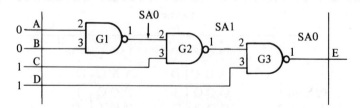

FIGURE 8-4 Detection of fault G1-1 SAØ.

8.3 Digital Test Systems · 155

With reference to Figure 8-5, the circuit and patterns illustrate the clock sequence. The fault is "exposed" in pattern 1 but is not detected until pattern 6. In this case, the trail is clocked through the register and exists from G1 to pin 1 to pin E at the moment that pattern 6 is clocked. If pins A and B were set to logic 1 in pattern 2, the trail would exist only in the register, but this fault would still be detected. Locating the fault involves tracking the progress of the failed node trail from the output to the input. In the combinatorial circuit of Figure 8-4, the probe would determine that pins A and B pass and that G1 pin 1, G2 pin 1, and G3 pin 1 fail. G1 is a device having good inputs and failed outputs and is therefore the source of the fault. In the sequential example, the entire pattern set must be repeated on each node. If any pattern fails, the node fails; otherwise the process is the same. Specific identification of the fault to the maximum possible degree will be detailed in Sec. 8-3, along with other fault isolation techniques.

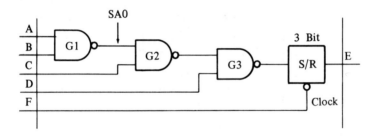

Pattern	A	B	C	D	E	F
1	0	0	1	1	1	1
2					0	1
3					1	1
4					0	1
5					1	1
6					0	0

FIGURE 8-5 Illustration of clock sequence.

8.3 Digital Test Systems

Tests of printed-circuit boards with standard test equipment is time-consuming and awkward on the laboratory bench. Because each board has many inputs and outputs, it requires either a mockup of the entire system for which the board is designed or a uniquely designed test fixture containing switches and lights. While sometimes acceptable in a design phase, this type of equipment is unsuitable for a production environment having many board types and large volumes of each. The programmable digital test system has been developed to meet digital requirements. A typical system is shown in Figure 8-6.

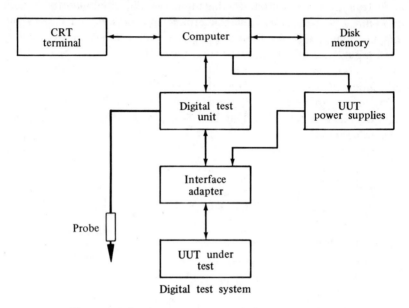

FIGURE 8-6   A programmable digital test system.

The digital test unit (DTU) is a stimulus–response mechanism that supplies input patterns to the unit under test (UUT) and reads output patterns from it. The interface adaptor provides a means of connecting the circuitry necessary to operate the UUT, such as pull-up resistors. UUT power supplies, which may or may not be controlled by the computer, provide various voltages for the UUT. The patterns supplied to the UUT are stored as a disc memory data file and are brought into the computer and sent to the DTU when the test is executed. They include both the stimulus and the expected failed response. The DTU generates a programmed delay after each stimulus pattern, after which UUT output states are sensed and compared with the expected response. A mismatch indicates a test failure. The DTU interrupts the computer, causing it to branch to a fault location routine, which analyzes the failed UUT output pattern and issues instruction through the CRT for the operator to probe internal board nodes. When the probe has been correctly placed, a signal from the operator to the system causes the entire test to be recycled. Probing proceeds along the fault trail until the source of the failure is reached.

A digital test unit, shown in Figure 8-7, is a chassis containing a number of PC cards having a maximum of several hundred driver-sensor pins. Each pin can be designated as an input or output (as referenced to the UUT) under program control. The chassis contains a control card and a backplane, which routes patterns to the appropriate driver-sensor card. Data are stored in the cards until all are loaded. (This may require several computer cycles since there are hundreds of bits and since the computer has 16-bit words.) A clock line is pulled, and the entire digital pattern is simultaneously clocked onto the UUT input pins. After a preprogrammed delay, which enables the UUT to settle to a static state, the UUT outputs are clocked into the result register. Should the result differ from the data expected, a "Fail" bit is set. This bit interrupts the computer, causing it to branch to a failure-processing routine.

A block diagram of a one-pin driver-sensor circuit is shown in Figure 8-8. It functions as either a driver or as a sensor, depending upon whether S1 is closed or open. The "load data" line is common to

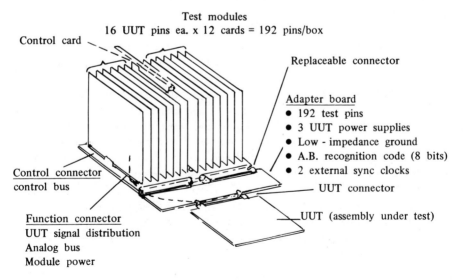

FIGURE 8-7 A digital test unit.

158 · Testing of Digital and Printed-Circuit Boards

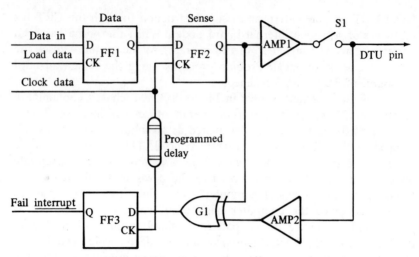

FIGURE 8-8   Driver-sensor arrangement.

all driver-sensors on a card and clocks in the desired data bit, an output state, or an expected input state, to FF1. When all of the data flip-flops are loaded, a common "clock data" line is pulled, and FF2 changes on all circuits to the desired bit and causes the driver amplifier to slew appropriately. If S1 is closed, the signal travels to the UUT. An incoming signal is routed to G1, an exclusive-OR gate which produces a logic 1 if the incoming bit differs from the signal expected in FF2. The programmed delay allows the UUT to settle to a static state. After it times out, FF3 is clocked and creates an interrupt if a failure has occurred.

Adding programmable slew limits to the driver amplifier and a dual comparator circuit to the sense circuit enables the arrangement to test a variety of logic families. Drive and sense are programmed independently. In the case of TTL, driver levels would be set to the minimum device input voltages of 0.8 volt for low and 2.4 volts for high. The sensor would be programmed to sense nominal output levels such as 0.4 volt and 4.8 volts. When references for both high and low are used, a logic-high minimum is established, below which a pin fails, and a logic-low maximum, above which a pin fails. Figure 8-9 shows the regions in which the window comparator causes failure. This circuit is able to catch "stuck-in-the-middle" failures.

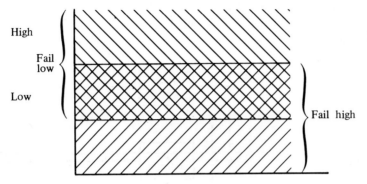

FIGURE 8-9 Window comparison diagram.

8.4 Probe Hardware

Basic requirements for probing internal nodes on a PC card are an ability to:

1. Read the node state for each pattern.
2. Sense an open-circuit condition.
3. Protect against an overvoltage condition.
4. Provide a variable reference for any logic family.

A comparatively simple circuit which accomplishes these functions is shown in Figure 8-10. This is a computer-guided probe. The probe input signal goes to a comparator, where it is compared with a reference from a DAC. The reference is set halfway between the logic 1 and logic 0 states of the logic family being tested. The first buffer amplifier, located in the probe, transmits the logic activity of the probe to the second buffer, where the signal is analyzed. The probe electronically analyzes the signal and sets bits indicating the status seen by the probe.

The static state bit is set to the state of the node at the end of the DTU delay time. The pulse state bit is set if two or more transitions occurred during the time between the input patterns and the delay time. The "wiggle" bit is set if a state change occurs and is used to determine whether a node is stuck for all patterns applied. The short pulse-detect indicates that spikes or pulses shorter than a given threshold, typically 50 μsec, exist on the line. This enables the node response to be characterized if pulses or glitches are necessary for correct performance.

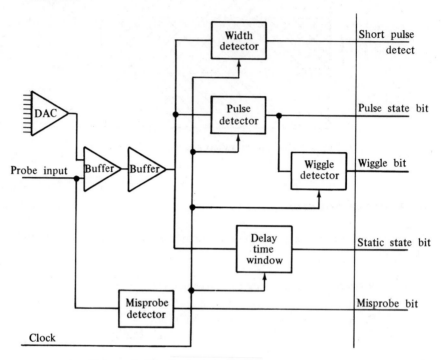

FIGURE 8-10 Probe configuration block diagram.

9.
TEST MODELS OF PRINTED-CIRCUIT CARDS

9.1 Primitive Elements

Complex digital-logic functions can be realized by interconnecting simple functions. Any logic function can be configured from NAND or NOR functions. In order to make the task of modeling a PC card manageable, logic simulators call upon many commercial digital IC devices, which can range from a quad NAND gate to an arithmetic logic unit (ALU) or even larger. By configuring these commercial devices, a model of an entire circuit board is built by naming individual devices and specifying their interconnections in a descriptive syntax, which is translated into a machine-readable data base. This data base is then used to perform logic calculations. Models are built from *primitive elements,* which are simple circuits known to the simulator in terms of their truth tables. A list of primitive elements is given in Table 9-1.

The basic elements can be used to construct most circuits in the logic simulator library. Although any circuit can be modeled with a NAND-equivalent, a variety of basic combinatorial elements is implemented for convenience and simulation speed. Memory elements have the feature (except for the D latch) of variable storage and avoid large computer memory requirements. Thus, a 1024-bit shift register

Table 9-1

PRIMITIVE ELEMENTS

Elements	Category
AND, OR, NAND, NOR, XOR, INVERTER, AMPLIFIER	Basic Elements
D-LATCH, SHIFT REGISTER, ROM, RAM	Memory Elements
CONNECTOR	Special Element
DELAY, INSERTING DELAY, MACRO DELAY	Timing Elements

requires 1024 bits of storage plus a small overhead. This is a great saving over a NAND-equivalent simulation.

Since basic models such as flip-flops and latches are repeated many times on a PC board, run time of a board model is drastically influenced by their complexity. While memory could be modeled of NAND-equivalents, core storage and computing times for this type of model are extremely high. Memory primitives are designed to functionally model operation of ROM's, RAM's, shift registers, and latches using a minimum of computer memory for model storage. Memory element size is variable according to the number of specified bits for ROM's, RAM's, and shift registers. Faults are not inserted in memory elements; techniques for developing patterns to test these elements will be detailed subsequently. Timing elements are of two types: micro and macro delays. The micro delay is used to simulate circuits in which propagation delay of a device is vital to its correct functioning. Figure 9-1 shows an example.

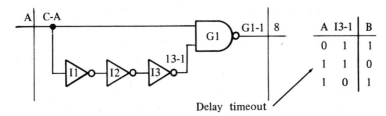

FIGURE 9-1   Example of a micro delay.

This circuit produces a negative pulse on the leading edge of an input pulse having a width equal to the propagation delay of the inverters. Since simulation is a mathematical process, provisions must be made for computing delays if they affect the operation of the circuit. The logic simulator is a zero-delay simulator, which means that all computations occur as if no propagation delays existed. The equivalent circuit is shown in Figure 9-2. If we assume that there is no delay, the circuit model appears not to change. (This is not strictly true, as will be discussed in Sec. 9-3.) The model is augmented with a micro delay to produce performance more comparable to the action of a real circuit.

FIGURE 9-2   A zero-delay simulator equivalent circuit.

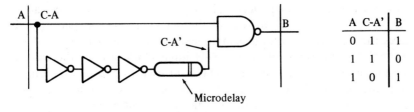

FIGURE 9-3   Example of micro delay.

164 · Test Models of Printed-Circuit Cards

Use of a micro delay (Figure 9-3) causes the logic simulator to calculate all changes in the circuit, time-out the delay, and then compute any subsequent changes. The result is the correct logical 1-0-1 transition. Micro delays are elements that alter the computational sequence; they have no real-time significance and may be thought of as unit delays. Macro delays are elements used to characterize the action of a long delay occurring as a result of a timing element, such as a one-shot. In the case of a long delay, the DTU measures an intermediate result, which occurs prior to the timing-out of the element. This occurs when the delay element is longer than the minimum interval at which the DTU is capable of reading a pattern. Depending upon the circuitry used, this time varies from 100 nsec to 5 μsec. Figure 9-4 shows the timing relationship.

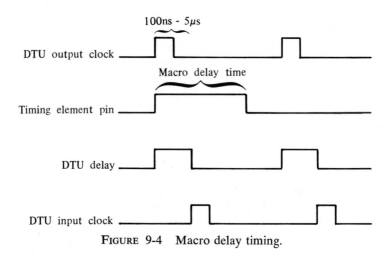

FIGURE 9-4   Macro delay timing.

The macro delay causes the simulator to produce two output patterns for a single input pattern. The first pattern indicates the state of the output prior to the time-out of the delay; the second pattern indicates the state of the output subsequent to the time-out of the delay. As indicated in Figure 9-4, the DTU input clock strobes the state of the timing element pin into the DTU 100 nsec to 5 μsec (or longer) after the output clock changes the state of the PCB input pins. The DTU output clock is then strobed again with an identical input pattern, and after a suitable delay, another pattern from the PCB output is clocked in. The second pattern contains the output states of the PCB after the macro delay has timed-out. The macro delay is thus used (and only) when the UUT response is slower than the minimum DTU response. Within the tolerance limits of DTU timing, it provides a measurement of the performance of a one-shot.

The *connector* is an element used to define the inputs and outputs of the device. A quad NAND gate having the connector configuration of a 7400 device is diagrammed in Figure 9-5. It provides the facility to identify device pins using numbers designated by the manufacturer.

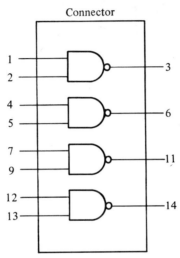

FIGURE 9-5 Diagram of 7400 quad NAND gate.

Connectors also determine "levels" of logic. If two devices are mounted on a PC board, each has a connector, and the board has a connector which interfaces them to the DTU. Level Ø modeling consists of interconnections at the connector of the PCB. Connections at the device connector are level 1 (Figure 9-6). Should a device be modeled from other models, a level 2 exists. Referring to Figure 9-7, level 2 is the boundary of the 7400 devices used to model the device U3. Connectors are the boundary for fault insertion. Faults are normally inserted at levels Ø and logic 1. It is possible to declare an "internal fault" option and insert at level 2. For modeling circuits, levels can be nested; however, faults can only be inserted in levels Ø through 2.

166 · Test Models of Printed-Circuit Cards

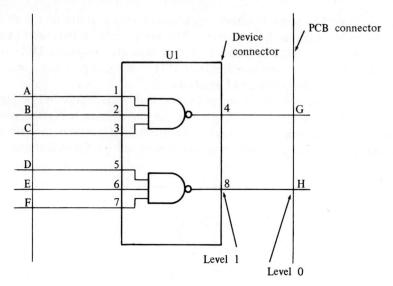

FIGURE 9-6 Example of level 1.

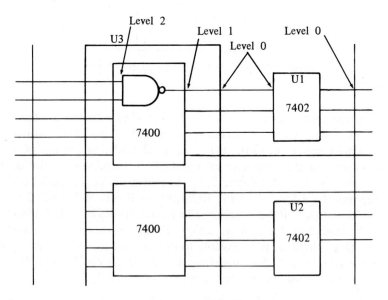

FIGURE 9-7 Example of level 2.

9.2 Macro Characterization

Complex logic configurations will frequently be encountered and are used repeatedly in commercial devices, the best example of which is the J-K flip-flop. Flip-flops and latches are difficult devices to model; nevertheless, they are basic building blocks for all sequential circuits. A circuit not labeled as a commercial device but used as a building block is called a *macro*. Macros for latches and flip-flops can be easily used to build all common sequential devices, including shift registers, counters, and memory elements. When using a macro, modeling a J-K flip-flop such as a 7473 is as simple as defining the manufacturer's connector pin designations. If the macro has been thoroughly tested, the model will work without further modification.

9.3 Tristate Modeling Techniques

Tristate devices are used to provide a convenient means of accessing a bus that is driven by more than one driver. The third state is a high impedance that will not interfere with another driver connected to the same node. Figure 9-8 shows a simple configuration. If control line B is low, receiver I3 responds to A; if control line D is low, it responds to B. If both lines B and D are low, the bus is driven by both drivers

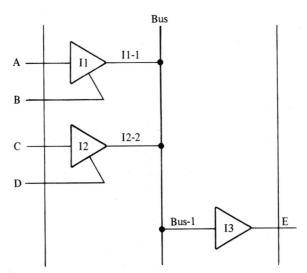

FIGURE 9-8 Simple tristate configuration.

and its state is indeterminate. If both are high, the bus is a high impedance and may float or be lightly biased by leakage from the receiver. Tristate elements require special consideration when using a simulator:

1. The high impedance state has no corresponding simulated state. It must be described using 1, Ø, and X. The X state, the third state calculated by the simulator, is defined as indeterminate. (The X state is not equivalent to the high impedance state because the bus would assume an X if any driver were X.) The function performed is illustrated in Figure 9-9 for a noninverting tristate buffer.

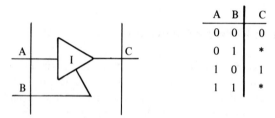

FIGURE 9-9   Tristate logic diagram.

2. When more than one driver controls the state of the bus, a logic equivalent of the wired connection is needed.
3. Bus lines may be used as both inputs and outputs. Since the simulator expects patterns at inputs and develops patterns for outputs, the connector description must be modified.

We can remove the X state produced when all drivers are disabled by creating a preferred condition for the bus. It can be biased high or low, but high is preferred to prevent reverse leakage currents from affecting its state. The high impedance state now becomes a logic 1 (Figure 9-10). A tristate buffer with a noninverting disable (high impedance on a 1 state) can be equivalently modeled with an OR gate. Different tristate elements will have different models, depending upon the bias state and whether the control or output function is inverting. Figure 9-11 illustrates the equivalent of a tristate inverter biased high. When we connect two drivers to the bus, its state is a logical function of their states. The circuit shown in Figure 9-12 assumes that two amplifiers are connected to a bus biased high.

9.3 Tristate Modeling Techniques · 169

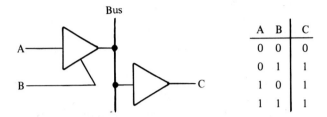

FIGURE 9-10 The high Z state becomes a logic 1.

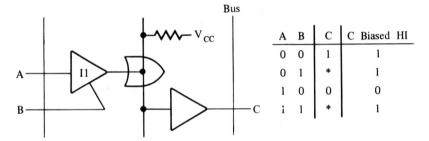

FIGURE 9-11 Equivalent of a tristate inverter biased high.

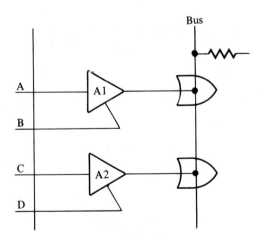

FIGURE 9-12 Two drivers connected to a bus biased high.

170 · Test Models of Printed-Circuit Cards

In the foregoing example, we assume that the tristate amplifiers, A1 and A2, go into the tristate mode when B is low and behave as normal buffer amplifiers when B is high or in an "enabled" state. This truth table has the equivalent circuit shown in Figure 9-13. When modeling a tristate element, take into account the bias applied, and develop the appropriate equivalent circuit. If the foregoing circuit were biased low, its equivalent would be a NOR function, as shown in Figure 9-14. The truth table in Figure 9-15 gives the states that will occur on the bus for all combinations of the A, B, C, and D states. A circuit that yields the desired states is depicted in Figure 9-16.

U1-A is modeled as a tristate buffer that is typically sold as a hex configuration (A through E). U2 is a similar device, the "A" driver of which drives the same bus as U1. The desired bus function is implemented by biasing the bus high and using an auxiliary circuit to produce the responses that are required. Biasing to a logic 1 enables us to test the driver element control line by applying a Ø and disabling the control as shown in Figure 9-17. The auxiliary circuit in Figure 9-16 serves to put an X on the bus if two drivers are simultaneously enabled and A1 on the bus is simultaneously disabled. The exclusive-OR gate, G4, enables G1 and G2 for any data from U1 or U2, as long as only *one* is enabled. If both are enabled, G4 becomes Ø, sensitizing the G3 to the bus. If both B and C are low (disabled), the bus is forced to a logical high. G4 and G5 are Ø, forcing NOR G3 to a 1. The wired-OR bus input is Ø Ø 1 from G1, G2, and G3, respectively, and the bus goes to a logical 1.

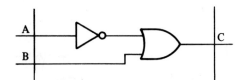

FIGURE 9-13 Equivalent circuit for truth table.

A	B	C Biased Low
0	0	1
0	1	0
1	0	0
1	1	0

FIGURE 9-14 Truth table for NOR function.

9.3 Tristate Modeling Techniques · 171

A1	A2		
A B	C D	BUS	
0 1	0 0	0	
1 1	0 0	1	A1 Enabled
0 1	1 0	0	A2 Disabled
1 1	1 0	1	
0 0	0 1	0	
1 0	0 1	0	A1 Disabled
0 0	1 1	1	A2 Enabled
1 0	1 1	1	
0 0	0 0	1	
1 0	0 0	1	A1 Disabled
0 0	1 0	1	A2 Disabled
1 0	1 0	1	
0 1	0 1	X	
1 1	0 1	X	A1 Enabled
0 1	1 1	X	A2 Enabled
1 1	1 1	X	

FIGURE 9-15 States for all combinations of A, B, C, and D states.

Elements G1 through G5 are not physically present on the board and thus cannot be probed. The guided probing algorithm has the ability to declare nodes "unprobable." When this is done, the next driving node is probed. Nodes on gates G1 pin 1 through G5 pin 1 are declared unprobable. The probe algorithm then probes the bus output and the connector pins B or D. The algorithm assumes unprobable nodes have failed, which causes the next probe instruction to point to the preceding driver. Obviously, if a mythical logic element fails, you have a "problem." The simulator inserts faults on the auxiliary circuit, which should be deleted to ease pattern generation.

The circuit of Figure 9-16 works for only two driving devices. An expansion to three devices is shown in Figure 9-18 and is a basic circuit that can be expanded further. Figure 9-19 shows the entire group of primitives organized as a macro. The inputs, labeled OUT A, OUT B, and OUT C are driven by the outputs of three tristate drivers, A, B, and C. CTL A, CTL B, and CTL C are connected to the control pins of the driver. The circuit is derived for tristate noninverting drivers,

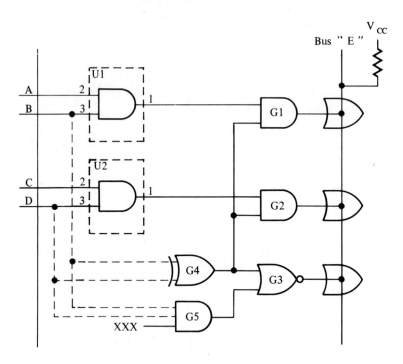

FIGURE 9-16 Circuit for obtaining the states noted in Figure 9-15.

which are enabled when high. The bus connector can be used as either an input or output to the board. The simulator recognizes only inputs or outputs. Recognize first that if there are drivers on the board, it may not be necessary to drive the bus with the DTU. The bus can then be considered as an additional output, as depicted in Figure 9-20.

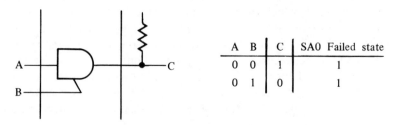

FIGURE 9-17 Test for control line.

9.3 Tristate Modeling Techniques · 173

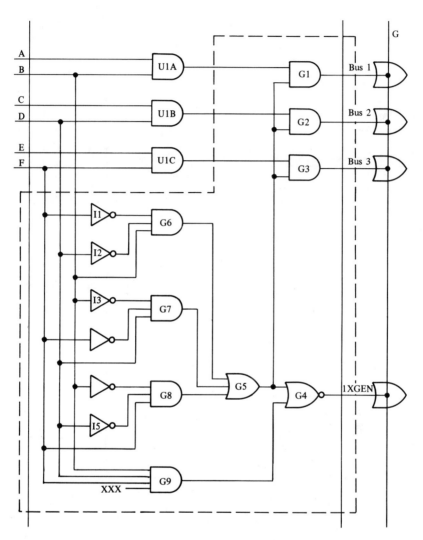

FIGURE 9-18 Tristate circuit for 3 noninverting controls, signals (TC3NCS).

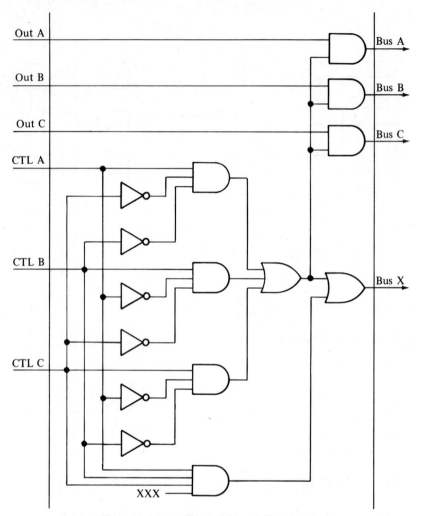

FIGURE 9-19 Three-driver TTL macro.

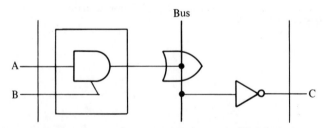

FIGURE 9-20 Bus can be considered as an additional output.

10 •

DIGITAL LOGIC SIMULATION IN TEST PROCEDURES

10.1 Simulation of a Model

Digital logic simulation is a process by which the response of a physical circuit to sets of stimuli can be determined by the use of a model instead of the actual implementation. The reasons for using a model are numerous:

1. The real time of the process is too rapid or too slow for purposes of practical measurement.
2. The cost of testing the real process is too expensive.
3. When variations of the real process are desired, it is too difficult to implement them physically.
4. When it is desired to examine the process with faults present, the quantity of units to be activated and tested is too large to be practical.

With the advent of the digital computer, it became practical to build "software" rather than physical models for use in simulation. The reasons for using simulation to develop tests of digital logic include all those listed above. Digital circuits respond very rapidly to stimulus and, in some cases, too fast for practical measurements. Breadboarding a real circuit is expensive, and it is easier to determine the performance of a model instead of that of the actual circuits. Modeling is essential, particularly when we wish to observe response in the presence of faults. In fact, it may be difficult to find a physical board that is fault-free and thereby serves as a means of specifying expected performance. A word of caution is necessary. The simulation results are only as good as the model, and it is impractical to build a model that duplicates the process in every detail. The user must know the model well enough to understand how it will respond to different stimuli and how to build it so that it provides expected results. A digital simulator can be thought of as calculating the response of logic elements. Consider the circuit in Figure 10-1.

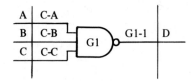

FIGURE 10-1 Concept of a digital simulator.

The simulator reads the states of input signals C-A, C-B, and C-C, and computes C-D according to a transform which is the truth table of the NAND-gate model. If inputs and outputs are selected from the set of binary values 1 and Ø, the process is termed two-value simulation. Unfortunately, logic simulation is not quite this simple, as can be seen from the circuit in Figure 10-2. Assume at a given time that signals C-A and C-B are at a logic 1. Then G1-1 will be a logic 1 if G2-1 is a logic Ø, or a logic Ø if G2-1 is a logic 1. However, if we calculate

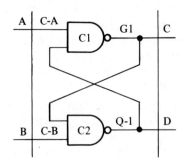

FIGURE 10-2 An example of difficulty in logic simulation.

the value of G2-1, we see that it is a function of G1-1. There are obviously two possible answers: G1-1 being logic Ø and G2-1 being logic 1, or G1-1 being logic 1 and G2-1 being a logic Ø. The fact that a signal can be either a Ø or 1 obviously complicates the simulation situation. The inclusion of a third-state X enables the simulator to correctly predict the response of a circuit during a state change, which could produce ambiguous results.

The X state means that the signal could be a Ø or 1 and is therefore indeterminate. Before progressing further, it may be useful to consider truth tables of a simple two-input NAND, with both 2 and 3 logic values, as depicted in Figure 10-3. It is useful to note that in some cases an X state on an input does not mean X on an input, and that in other cases it does. We can say that a gate is "sensitive" if the X progresses through it. A two-input NAND is sensitive if one input is 1; a two-input NOR gate is sensitive if one input is a Ø. In the case of a two-valued simulator, a logic Ø simply becomes a logic 1 and vice versa. What about the three-valued simulation? For the moment, let us assume a signal changing from Ø to 1 will have the states Ø-X-1. Similarly, a signal moving from a logic 1 to a logic Ø will have the states 1-X-Ø. With reference to Figure 10-4, assume that signal 1 changes

Two State				Three State		
Inputs		Outputs		Inputs		Outputs
A	B	C		A	B	C
0	0	1		0	0	1
0	1	1		0	1	1
1	0	1		1	0	1
1	1	0		1	1	0
				1	X	X
				0	X	1
				X	1	X
				X	0	1
				X	X	X

FIGURE 10-3 A two-input NAND gate and truth tables.

	T1	T2	T3	T4	T5
A	1	X	0	X	1
B	1	1	1	1	1
C	X	X	1	1	1
D	X	X	0	0	0

FIGURE 10-4 Transition table.

from a logic 1 to a logic Ø and then back to a logic 1. The following transition table can be derived, as shown in Figure 10-4.

The simulator would calculate the response as follows: at T1, C and D are X from the previous example. When A becomes an X, nothing changes since it could still be a 1. However, once that A becomes Ø, then C must be a 1 and, since B and A are both 1, D must be a Ø. Our simple latch circuit has become initialized. At this point, we must be aware of the order of calculation in the simulation process. At T3, D is Ø and A is also Ø. The next thing that happens is that A changes to X at T4. The first calculation that the simulator will make is the response of C. Since D is still Ø, C cannot change. In consequence, C remains the same at D as it was in T3, and there is no need for further calculation later. At T5, signal A becomes a 1, but again no changes result. At T4, A could be either a Ø or a 1, so it is not surprising that C does not change.

It is fundamentally important to realize the order in which the calculations propagate and, therefore, the way in which signal states are calculated. Reviewing the process between steps T2 and T4 may be important. Between T2 and T3, A was the first signal to change, moving from an X and a 0. This then, causes C to change from an X to a 1. Since C changes, gates following must be evaluated, with the result that D changes from an X to 0. Since D changes, it is again necessary to recalculate C. But in this case, with A and D being Ø, C did not change. Hence, the calculation was terminated. It is important to recognize why C and D are not X at T5. From this example comes the first fundamental of simulation. When a change is made to an input, calculation of the change propagates only until responses do not change, and calculation is then terminated. Note that in the transition between T3 and T4, the value of D at time T3 was used in the initial calculation.

Another important concept should be observed from the foregoing example. At T5, C and D have assumed values different from those derived from the static case. This is because previous states (even though they were unknown) and transitions were evaluated and the latency of their behavior determined.

10.2 Races, Critical Races, and Hazards

A race occurs when two or more signals at the input to a logic circuit occur simultaneously. If the order in which two signals reach an element affects its output, a critical race occurs. An example of a critical race is shown in Figure 10-5. The static condition of C remains unless B reaches 1 before A reaches Ø. In this case, the gate will emit a negative pulse, called a hazard. In a static test, the test equipment looks at the circuit output long after the pulse occurs; the hazard has no effect unless it changes the state of a memory element. A test pattern set must avoid critical races and hazards if consistent results are to be obtained during test procedures.

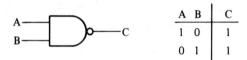

A	B	C
1	0	1
0	1	1

FIGURE 10-5 Example of a critical race.

When we observe propagation of a race using a three-state simulation, we get the transitions shown in Figure 10-6. The 1-X-1 sequence indicates that, during the transition time, a Ø state *may* have occurred. In a physical circuit, signals A and B will not arrive simultaneously, and C may or may not momentarily change, producing a spike or glitch —a hazard. All possible hazards are exposed by a three-state simulator through calculations of 1-X-1 or Ø-X-Ø sequences. The example shown in Figure 10-7 depicts the effect of X on a simple latch. An X response from a device will propagate through all of the logic that can be af-

A	B	C
0	1	1
X	X	X
1	0	1

FIGURE 10-6 Transitions of a race in three-state simulation.

fected by the indeterminate node. This point is illustrated by the example in Figure 10-8.

Signal C-A disables G1, forcing E to 1; signal C-D enables G2, and F assumes the X state of FF1 $\bar{Q}$, which was cleared by the race of signals C-B and C-C. The X state produced on pin F must be cleared by again making the state of FF1 determinate. The role of X is to place the burden of reinitializing the flip-flop on the test programmer. In some circuits, correct operation occurs because propagation delays resolve a race. Referring to Figure 10-9, a pulse is generated. The propagation delay effects can be simulated with a micro delay element, which has

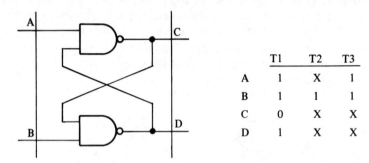

	T1	T2	T3
A	1	X	1
B	1	1	1
C	0	X	X
D	1	X	X

FIGURE 10-7   Effect of X on a simple latch.

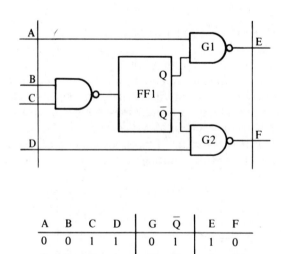

A	B	C	D	G	$\bar{Q}$	E	F
0	0	1	1	0	1	1	0
0	1	0	1	X	X	1	X

FIGURE 10-8   Propagation of an X response.

10.2 Races, Critical Races, and Hazards

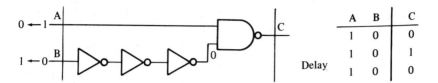

FIGURE 10-9 Race resolved by propagation delay in one line.

the property of modifying the sequence of computation. Observing the transition table for a delay in Figure 10-10, we see that a delay prevents the output from changing until a complete transition occurs at the input. We may think of the delay as altering the calculation sequence occurring as a result of a change.

	T1	T2	T3	T4	T5	T6	T7	T8	T9
A	0	X	1	1	1	X	0	0	0
B	0	0	0	X	1	1	1	X	0

FIGURE 10-10 Transition table for a delay.

The circuit model depicted in Figure 10-11 illustrates the use of a delay to resolve races, such as those that occur owing to the propagation delay in Figure 10-9. In this case, signal DEL-1 is delayed with respect to A. The transition table is:

	T1	T2	T3	T4	T5
C-A	Ø	X	1	1	1
C-B	1	X	Ø	Ø	Ø
DEL-1	1	1	1	X	Ø
C-C	1	X	Ø	X	1

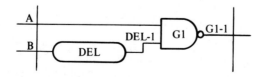

FIGURE 10-11 Use of a delay to resolve a race.

A negative pulse is thus produced by signal G1-1. It is possible to say that by using delays, we would have a model that more accurately reflects circuit performance. In Figure 10-12, a transition from ØØ to 11 will cause the model (and a physical circuit) to oscillate. Oscillations of a model are undesirable, because a simulator cannot track the infinite number of state changes caused by them. The desired result is achieved with the more idealized zero-delay model in which the X outputs propagate through the rest of the circuit, forcing the test designer to clear the race and reinitialize.

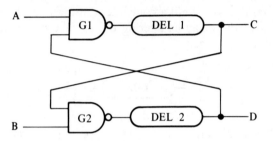

FIGURE 10-12   A model with the possibility of oscillation.

10.3 Model Verification

The objective of modeling is to create a software simulation of a device that accurately simulates the key parameters of interest. When a designer uses a device for a specific application, he knows what states it will assume and will try to ensure that the circuit responds as expected. When testing the board by developing a pattern set that attempts to detect all faults, device inputs can be entirely different. Illegal states and race conditions occur, particularly if care is not taken when patterns are developed. The logic state response of models must be absolutely identical to states of the physical circuit, or false faults will be detected. Nevertheless, building a model that performs properly for every possible combination of input states, statically and dynamically, is not always feasible. For example, decoder circuits from different manufacturers behave differently when "illegal" static states are supplied to the inputs.

Besides being logically correct, the most important requirement of a model is to ensure that X states are produced whenever the performance of the physical device is uncertain, whether owing to races, oscillation, hazards, or differences in response among manufacturers. There are basically four steps to ensure that a device model performs correctly:

1. Make certain that *static performance* is consistent with the manufacturer's data sheet (or, for a macro, the desired performance).
2. Make certain that *dynamic performance* (i.e., the transition response to all possible input race conditions) agrees with the manufacturer's data sheet (which is typically deficient in this respect) or known device performance, or is indeterminate (X states).
3. Make certain that all possible feedback configurations of the model do not oscillate. The model should assume an X state if the physical device oscillates.

The example in Figure 10-13 illustrates the process of creating a good model. A quad version of this circuit is the SN7475. An easy way to model the device is to construct an RS latch macro, verify it, and use it four times with the appropriate connector. Looking at the truth table of the configuration, the first three states can be obtained from the manufacturer's data sheet. The fourth state is 11. If the circuit reaches 11 from 1Ø or from Ø1, it will retain the previous state. Going from ØØ to 11 causes a race, making the output indeterminate. Remembering we have no control over how this circuit will be used or how patterns that test or pass through it will be generated, we therefore assume the race to be among the possibilities to get to 11 and use the XX output as the desired response.

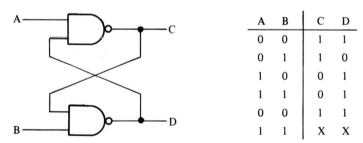

FIGURE 10-13 Process of developing a good model.

Commercial circuits have been designed to prevent races and oscillation, usually by adding considerable complexity. The J-K flip-flop

184 · Digital Logic Simulation in Test Procedures

depicted in Figure 10-14 provides a good example of the tests that we must make on a model to ensure proper performance. The patterns in Figure 10-14 will verify operation of the model and detect all faults. The patterns noted with (*) have races that are noncritical. The next step is to add combinations that are *critical* races to ensure that the model performs properly. If there is doubt that a race is noncritical, it should be tested. Inputs are paired off to test for races. For noncritical races, the model should go to the correct known state. For critical races, the output should go to X. Figure 10-15 gives patterns for race-checking a J-K flip-flop.

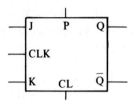

Static Patterns

	P	C	J	K	CLK	Q	$\overline{Q}$	Path/State
1	1	1	1	1	1	X	X	
2	0	1	1	1	1	0	1	QP/Q
3	1	1	1	1	0	1	0	QP/1
*4	1	1	0	1	1	1	0	QJ/Q
5	1	1	0	1	0	0	1	
*6	1	1	1	1	1	0	1	QJ/1
7	1	1	1	1	1	0	1	
*8	1	1	1	0	1	1	0	QK/1
9	1	1	1	0	0	1	0	
*10	1	1	1	1	1	1	0	QK/0
11	1	1	1	1	0	0	1	
12	1	1	1	1	1	0	1	QC/0
13	1	0	1	1	1	0	1	
14	1	1	1	1	0	1	0	QC/1

FIGURE 10-14 Characteristics of the J-K flip-flop.

Race	P	C	J	K	CLK	Q	$\bar{Q}$
	1	1	1	1	1	X	X
PC/0	0	0	1	1	1	X	X
PC/1	1	1	1	1	1	X	X
P CLK/0	0	1	1	1	0	0	1
P CLK/1	1	1	1	1	1	0	1
C CLK/0	1	0	1	1	0	1	0
C CLK/1	1	1	1	1	1	1	0
*J-0 CLK/0	1	1	0	1	0	0	1
J-1 CLK/1	1	1	1	0	1	0	1
	1	1	1	0	0	1	0
K-1 CLK/0	1	1	1	0	1	1	0
*	1	1	1	1	0	1	0
J-1 CLK/0	1	1	0	0	1	1	0
*	1	1	1	0	0	1	0
K-1 CLK/1	1	1	0	1	1	1	0
	1	1	0	1	0	0	1
J-1 K1 CLK/0	1	1	0	0	1	0	1
	1	1	1	1	0	0	1
J0 K0 CLK/0	1	1	1	1	1	0	1
	1	1	0	0	0	1	0

Rows PC/0 through J-1 CLK/1 bracketed: No oscillations

FIGURE 10-15 Patterns for race-checking a J-K flip-flop.

10.4 Board Modeling

When the device models to be used for a PC card circuit have been verified, the circuit is coded in the description language for the simulator and is ready for simulation, except for a final check for race conditions. Races come from two sources: parallel logic paths, which converge on a single device (Figure 10-16), and feedback paths (Figure 10-17). The astute observer will immediately pick out the fact that the circuit in Figure 10-14 is redundant and does nothing when the flip-flop is enabled, as is shown in the transition table. However, the signals that reach G2 arrive by two paths and thus constitute the same race condition shown for a two-input NAND gate. A circuit such as this may not

186 · Digital Logic Simulation in Test Procedures

be apparent to a designer who configured it using the flip-flop for a different purpose. The unanticipated states created for testing enable the redundant path. Note that the hazard occurs when only one input pin is pulsed. On a real board, the hazard may never occur, because the relatively large propagation delay of the flip-flop resolves the race to a Ø1-11-1Ø sequence on G2, which does not produce a pulse. The appropriate action when modeling the board is to insert a micro delay after the flip-flop.

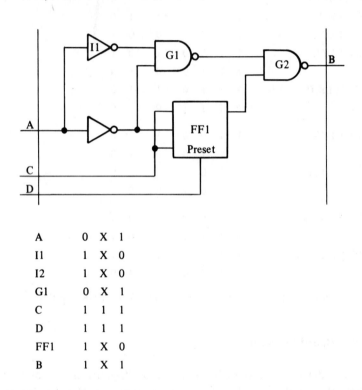

A	0	X	1
I1	1	X	0
I2	1	X	0
G1	0	X	1
C	1	1	1
D	1	1	1
FF1	1	X	0
B	1	X	1

FIGURE 10-16 A race situation.

The feedback in the circuit of Figure 10-17 will cause a latch-up in an X state whenever D changes while B is in a 1 state, although the propagation delay of a physical circuit would prevent this. If this configuration were seen in a schematic, the test engineer should add a delay. Circuit operation proceeds as in Figure 10-18.

10.4 Board Modeling · 187

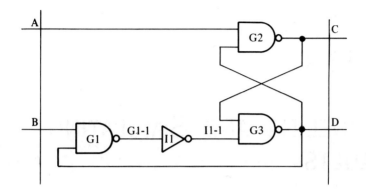

Software/No delay

A B	G1-1	I1-1	D C
0 0	1	0	1 1
0 X	X	X	X X
0 1	X	X	X X

FIGURE 10-17 Another race situation.

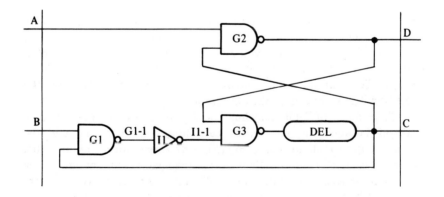

AB	G1-1	I1-1	D C
00	1	0	1 1
0X	X	X	1 1
01	0	1	1 1
01	0	1	0 1

FIGURE 10-18 Circuit action in the arrangement of Figure 10-17, with delay.

11 ·

MODELING AND SIMULATING FAULTS

11.1 General Considerations

Previous discussion has been concerned with simulation of a normally functioning circuit. However, the same simulation can be employed for faulty circuits. The reason for simulation of faulty circuits is to determine if faults are detected when we apply patterns. Detection, therefore, implies that the faulty circuit will behave differently from the good circuit as viewed from the output pins.

A problem of fault modeling should be viewed in the light of a trade-off between the number of calculations to be made and the quality of the simulation, as compared with "real world" failures that occur. It is possible to simulate every imaginable fault at great cost without proportionately increasing test quality and cost savings. Consider what happens during simulation. For each fault and each pattern, all node states of a good and bad board are computed. A board containing 1000 faults would require about 10^6 calculations of all node states to evaluate 1000 patterns, if all faults were simulated. If each evaluation required 10 msec, 1000 seconds would be required. Doubling the number of faults would double the time; however, new patterns would be added to detect new faults. If 1000 new patterns were required, the total compiled time would quadruple.

The faults being discussed are actually alterations of the circuit models being simulated. They closely represent some of the things that can occur with a physical circuit when it fails. The vast majority of digital circuit failures result from internal short or open circuits and are manifested as static "stuck-at" conditions. In forms of nonsaturated logic, such as MOS, resistive regions can be stressed and will change value. The result is a signal that reaches the correct logical value although at a great degradation of the rise-time specification. The same effect may occur if a pull-up resistor is missing or open. This type of failure can be converted to a stuck-at variety if the DTU samples the signal shortly after stimulus is applied, as illustrated in Figure 11-1. The DTU measures a logic Ø or stuck-in-the-middle condition for the second waveform. If logic intervenes, the PC board output waveform may have sharp but delayed transitions, as shown in Figure 11-2. Restricting ourselves to static stuck-at-1 (SA1) and stuck-at-Ø (SAØ) faults, we can define several classes.

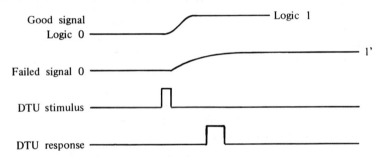

FIGURE 11-1 Analysis of signal with excessive rise time.

Node faults are SAØ or SA1 conditions on an entire circuit node. They can be represented by disconnecting the node from its driving device and connecting it to a permanent logic 1 or Ø condition, which we will designate V_{cc} or Gnd, respectively. A pin fault entails a SA1 or SAØ condition caused by the breaking of an IC pin or solder joint, a bond, or a PC card trace. The fault "splits" the node; the portion connected to a driver stays at the correct state, and the portion connected to the receiver drifts to the opposite state. In TTL, the receiver will drift to a logic 1; consequently, a SAØ pin fault would not occur for this logic family. If the drift of the open node is to a logic Ø, the SA1 pin fault does not exist. (See Figures 11-3 and 11-4.)

Input–Output faults are node faults on the input connector, and pin faults on the output connector. I/O faults are equivalent to other

190 · Modeling and Simulating Faults

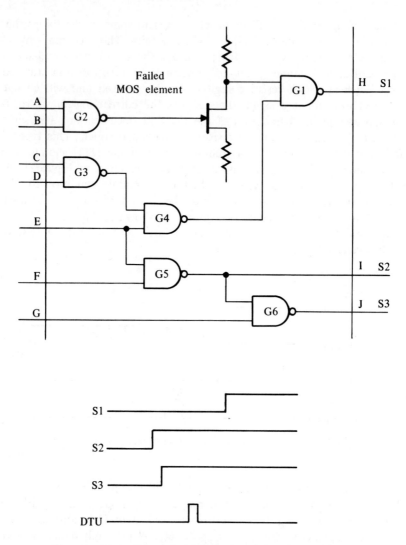

FIGURE 11-2 Example of sharp but delayed transition.

node and pin faults and will test most of the internal nodes if detected. Figure 11-5 shows I/O faults for a simple circuit. The pattern set shown moves all nodes and detects all I/O faults, but it misses G1-1 SA1. Specifying node faults would prevent this occurrence by requiring the test engineer to generate an additional pattern that forces G1 PIN1 to a Ø while C PINC is a 1.

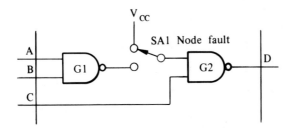

FIGURE 11-3   G1 PIN1 SA1 node fault.

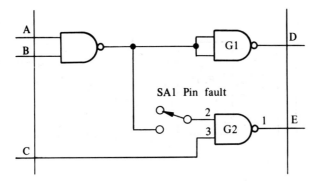

FIGURE 11-4   G2 PIN2 SA1 pin fault.

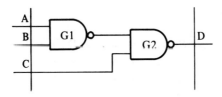

A	B	C	D	IO Faults detected
0	1	1	0	C Pin D SA1 C Pin A SA1
				C Pin B SA0 C Pin C SA0
1	0	1	0	C Pin A SA0 C Pin B SA1
1	0	0	1	C Pin SA0 C Pin C SA1

FIGURE 11-5   I/O faults.

192 · Modeling and Simulating Faults

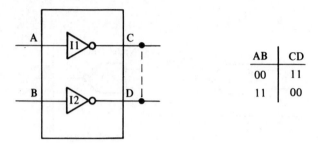

FIGURE 11-6 Example of shorted fault.

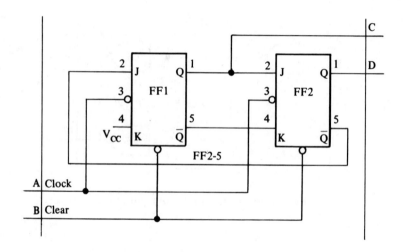

Pattern	A	B	C	D	$\overline{Q}$
1.	1	0	0	0	1
2.	1	1	0	0	1
3.	0	1	1	0	1
4.	1	1	1	0	1
5.	0	1	1	1	0
6.	1	1	1	1	0
7.	0	1	0	0	1
8.	1	1	0	0	1
9.	0	1	1	0	1

FIGURE 11-7 Portion of a 4-bit divide-by-12 counter.

Shorted Faults are caused by solder blobs or other factors connecting two pins, neither of which is at ground or V_{CC}. The shorted fault would not be detected by the pattern set shown. In the case of a pattern set that does not detect a shorted fault, the shorted pins will drift high, low, or in between. This condition is equivalent to one or two SA1 or SAØ node or pin faults. (See Figure 11-6.) Since our objective is to replace defective devices, it is usually sufficient to consider faults manifesting themselves as device pins SA1 or SAØ; however, a complex device may give incorrect results due to an *internal fault*, even if all outputs move to both states.

Figure 11-7 shows a portion of a 4-bit divide-by-12 counter. Node and pin faults involve A, B, C, and D stuck-at-1 and stuck-at-Ø. All node faults will be detected by patterns 1–5. However, pattern 6 will fail if signal FF2-5 is SA1, since it is an internal state not moved to Ø by the clock or preset. In Figure 11-8, two faults are illustrated. A node fault and a pin (input pin) fault are depicted. The major difference between them is that the node fault causes the whole node to assume the fault state. In the case of a pin fault, only the following gate assumes the faulty condition—not the node. Therefore, the presence of the pin fault illustrated in Figure 11-8 would not affect the output of G1 or A2.

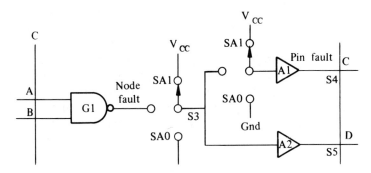

FIGURE 11-8   Illustration of two faults.

11.2   Detection of Faults

To illustrate detection of faults, consider the circuit shown in Figure 11-8 with signals C-A and C-B set to 1. With reference to Figure 11-9, for a single test (S1 = 1, S2 = 2) both the SA1 node and SA1 pin fault are detected. Note that the outputs (A1-1 and A2-1) with these faults are different from outputs of the good circuit. The foregoing

194 · Modeling and Simulating Faults

circuit was illustrated with just two faults (1 node and 1 pin). Since there are five nodes on the circuit, there are five places that node faults could be inserted. Similarly, there are six places that pin faults could be inserted (two inputs to G1, the input to A1, the input to A2, and at output connector pins C and D).

	C-A	C-B	A1-1	A2-1	
Good circuit	1	1	0	0	
Node fault SA1	1	1	1	1	SA0 Node fault detected
Node fault SA0	1	1	0	0	
Pin fault SA1	1	1	1	0	SA1 Pin fault detected
Pin fault SA0	1	1	0	0	

FIGURE 11-9 Fault patterns.

Obviously, on a large circuit board, there would be many faults to simulate. Fortunately, many of the faults are logically equivalent and can be simulated as one. For example, consider the situation shown in Figure 11-10. Considering only node faults, we can write the groups depicted in Figure 11-11. In turn, these six node faults can be combined into two groups. Note that simulation of any fault will produce an output identical to that of any other fault in the group. Therefore, for simulation purposes, a single fault is selected from each group for simulation. In the foregoing example, it can also be shown that all of the pin faults could be placed in the two groups. In fact, all 12 faults (6 node and 6 pin) can be combined into two groups for simulation, resulting in a great saving of computer time. When faults are simulated, each fault (or group) is simulated separately. It would be too

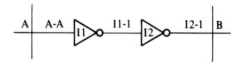

FIGURE 11-10 Circuit for group fault analysis.

                A-A (SA1) = I1-1 (SA0) = I2-1 (SA1) = Group 1
and   A-A (SA0) = I1-1 (SA1) = I2-1 (SA0) = Group 2

FIGURE 11-11 Example of fault grouping.

time-consuming and unnecessary to consider the effects of combination of two faults occurring simultaneously. However, advantage can be taken of the fact that the simulator is capable of simulating more than one group of faults at a time.

Consider the circuit shown in Figure 11-12. Faults associated with C-A, C-C, and I1 can only affect output pin C. Similarly, faults associated with C-B, C-D, and I2 can only affect output pin D. Faults associated with output pin C cannot interact with faults associated with D. Therefore, for simulation purposes, these faults can be combined. Output pin states relate only to special portions of the circuit. Any fault detected on pin C can be related to C-A, I1, and I2 only—*not* C-B, I2, and C-C. Similar reasoning applies to pin D. This combining of non-interactive faults again results in a significant saving in simulation time. Most faults will result in failed outputs that are distinctively different from the good board. However, two exceptions should be noted.

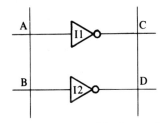

FIGURE 11-12   More than one group of faults can be simulated at a time.

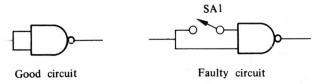

FIGURE 11-13   A redundant configuration.

Consider the circuit depicted in Figure 11-13. A truth table derived for both the good and the faulty circuit is the same. The stuck-at-1 pin fault cannot be detected because the circuit is redundant; a simple inverter would have been adequate instead of a two-input NAND gate. Nevertheless, redundant circuitry does exist, and it is important to recognize it. Otherwise, the designer will become frustrated trying to develop tests for "undetectable faults."

11.3 Levels of Fault Insertion

It is useful to think of faults as being inserted after connectors. Primitive elements can be thought of as having an imaginary connector. In Figure 11-14, we would insert a pin fault on U1 pin 1 by going to pin 2 of the connector and placing a fault on the next element input, G1 pin 2. The fault would be designated as U1 pin 1 SA1. The site of insertion, as far as the simulator is concerned, is identical to the site of G2 pin 2 SA1, an equivalent fault. G2 pin 2 SA1 is behind the connector of U2 and is thus a level 1 fault. In Figure 11-15, two chips are

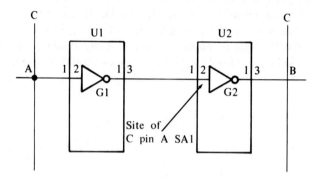

FIGURE 11-14 Example of fault insertion.

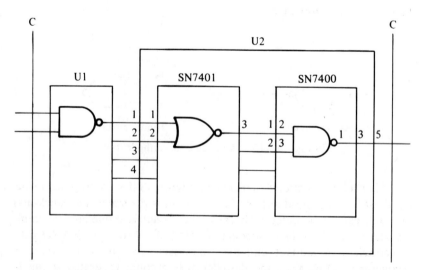

FIGURE 11-15 Hybrid device formed from two chips on a substrate.

mounted on a substrate to form a hybrid device, designated U2, which is mounted on a PC board. This structure contains level Ø I/O faults, node and pin faults on connector C and on U1 and U2. Level 1 faults can be inserted on the pins of the chips. Level 2 faults could be inserted on the pins of the primitives inside the chips. These are internal faults. Simulation of faults at level 2 or lower is impractical. In most MSI circuits, such as shift registers and counters, these are equivalent to lower-level faults or are detected along with them. Only those faults on nodes completely internal to the device are unique, and these typically have a lower failure rate than faults on the I/O pins. Shift registers and counters have "carry" nodes that are internal. However, faults on these nodes are propagated out if bits are shifted through a register or any functional model; faults may have no meaning. If all internal faults are simulated for a large board, computation time increases significantly. The fundamental question is whether the increased quality of the resulting pattern set justifies the added expense of simulating internal faults. Experience with test quality obtained by simulating level 1 node, pin, and I/O faults suggests that it isn't.

11.4 Fault Machines

The term *fault machine* is used to identify a logical model of a circuit having a single fault. (The good board model is called the *good machine*.) Any pattern applied to a fault machine will produce outputs characteristic of a faulty board. The board depicted in Figure 11-8 has a total of 22 I/O, node, and pin faults. A fault-inserting simulator creates 23 machines in the form of data files and computes the output of each for the pattern being simulated. As faults are detected, machines can be eliminated to reduce computation. Each machine is a model of the board containing a single fault.

11.5 Logically Equivalent Faults

The fact that many faults in a circuit are logically equivalent is important to using the simulation approach, because several equivalent faults can be represented by one machine for which computations are carried out. No information about the quality of the pattern set is lost, because the list of equivalent faults is tacked on to the failed response. An equivalent fault is defined as having the property that any pattern which detects it, detects the entire set of its equivalents. The simplest case is

198 · Modeling and Simulating Faults

FIGURE 11-16 An inverter for fault analysis.

the inverter depicted in Figure 11-16. I1 PIN2 SAØ is logically equivalent to I1 PIN1 SAT1. Equivalent faults have identical failed outputs.

A distinction should be drawn between equivalent faults and simultaneously detected faults. Simultaneously detected faults may have identical or different failed output patterns. In Figure 11-17, all responses for an 11-input pattern are given (if faults are omitted). A 1Ø output detects the faults of G1 PIN1 SAØ and G1 PIN2 SA1. Although simultaneously detected on the same pattern, these faults are *not* equivalent. When a ØØ input pattern is applied, a ØØ fault signature detects G1 PIN1 SAØ, and a different fault, G1 PIN3 SA1. Thus, the faults are not pattern-independent.

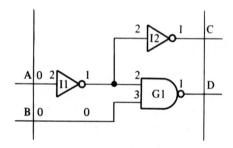

A	B	C	D			A	B	C	D	
1	1	1	1	Good response		0	0	0	1	
		1	0	G1 Pin1 SA0, G1 Pin2 SA1		0	0			G1 Pin1 SA0
										G1 Pin3 SA1

FIGURE 11-17 Equivalent versus simultaneously detected faults.

11.6 Fault Signatures

As noted previously, logic simulation provides a test response of both faulty and good boards. The faulty output for a given input pattern is called a *fault signature*. We can quickly locate a single fault that causes

11.6 Fault Signatures · 199

a signature by comparing the observed failed output with results of a previously run simulation. Unfortunately, two problems intervene: equivalent faults have identical signatures, giving several possible causes of a given output state, and multiple faults produce new, unsimulated signatures. In turn, the fault can be detected but not located. This situation is illustrated in Figure 11-18. If both FF1 PIN1 SA1 and FF1 PIN4 SAØ occurred simultaneously, the signature would be 1Ø. This is clearly a faulty response and would be detected. However, the signature would never have been simulated, since faults are only simulated singly.

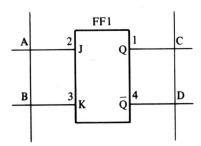

Fault	Good response	Signature
	CD	CD
FF1 Pin 1 SA1	01	11
FF1 Pin 1 SA0	10	00
FF1 Pin 4 SA1	10	11
FF1 Pin 4 SA0	01	00

FIGURE 11-18 Typical fault-signature testing problems.

When software logic simulators were first developed, a great deal of emphasis was placed upon the use of fault signatures as the sole fault-location mechanism. If a successful signature technique is implemented, the faulty IC can be replaced without further testing of internal node states. In military applications, where humidity seals, potting of components, and delicate boards are common, application of fault signatures has great appeal. The usual technique, if a guided probe algorithm is to be used, is to discard fault machines as soon as their faults are detected, reducing the number of machines for which calculations are performed on all subsequent patterns. A great saving in simulation time is thus achieved. Additional resolution of fault signatures is achieved if machines are not discarded. This mechanism is illustrated in Figure 11-19.

Although each signature contains two faults, one is common to both and is thus uniquely located. All of these fault machines were simulated twice. Had a single signature been used, G2 PIN1 SAØ and G2 PIN3 SA1 would have been discarded during simulation of Pattern 1. The retained fault technique is most effective if models are based on NAND-gate equivalents and internal faults are simulated. More equivalent faults are developed, and more patterns and signatures must be developed to achieve detection. The additional signatures provide a larger data base for the same minimized fault set and thus enable additional resolution of fault isolation. The problem with this approach is that computation increases as does the product of additional faults and patterns, and this is multiplied by the increased number of circuit nodes if NAND-equivalent modeling is employed. Moreover, all fault machines are retained for all patterns. Only large computers can handle this type of simulator, and very high operating costs result. Only certain military programs have effectively used this approach.

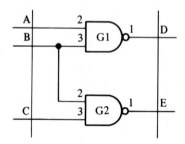

Pattern	A	B	C	DE	Signature	
1	0	1	0	11	1 0	G2 Pin 1 SA0, G2 Pin 3 SA1
2	1	0	1	11	1 0	G2 Pin 1 SA0, G2 Pin 2 SA1

FIGURE 11-19 Example of additional resolution of fault signatures.

11.7 Stop on First Fail

There are several subtle advantages to the approach of dropping fault machines upon first detecting a fault, called the *stop on first fail* method. Consider again the example shown in Figure 11-19. The "Run All Patterns" method would compute two fault machines for two patterns. If we drop any detected faults immediately, the second pattern computes only one machine, since G2 PIN1 SAØ has already been detected. Thus, we only compare a total of three machines. We also *know* that if G2 PIN1 SAØ has been eliminated on Pattern 1, it is not a possible fault on Pattern 2. If the DTU executes Pattern 2, we know that Pattern 1 has passed. We therefore know that Pattern 2 uniquely detects the fault G2 PIN2 SA1.

11.8 Multiple Fault Sites

Consider the circuit used as a device shown in Figure 11-20. Assuming that the likely cause of a pin fault is a broken bond and that both gates are on one chip, the effect of the fault U1 PIN2 SA1 is to open both G1 PIN3 and G2 PIN2. The simulator identifies U1 PIN2 as a fault. However, there are two sites at which it could be simulated: G1 PIN3 and G2 PIN2. Because the simulator treats faults singularly, the multiple effect caused by the bond is not correctly simulated. The simulator will pick one of the sites and will indicate detection when given an appropriate pattern; however, signatures will not correspond to actual faulty circuit performance. With reference to Figure 11-21, the pin fault U2 PIN2 now becomes equivalent to A1 PIN2. The total number of inserted pin faults is reduced, saving computation in circuits having many internal fanout points. This model is physically consistent because inputs and outputs of most commercial IC's are buffered.

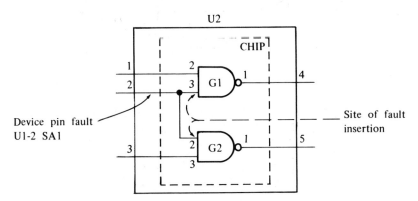

FIGURE 11-20 Circuit for fault analysis.

202 · Modeling and Simulating Faults

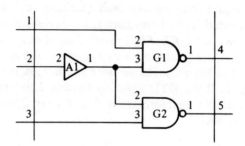

FIGURE 11-21   Example equivalent configuration.

11.9   Trade-offs on Selecting Faults

Pin faults are more numerous than node faults and impose an additional computational and pattern generation burden on the test programmer. However, there are benefits to be obtained by declaring them. They do occur in practice and should be detected. It is possible to miss them if patterns do not sensitize all paths to the output. In Figure 11-22, G2 PIN2 SA1 is not detected by the patterns shown, because G2 PIN2 is sensitive only when it is SA1. When developing a manual pattern set for a large board, computation time can be saved by declaring I/O

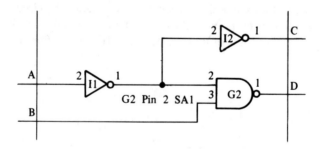

G2 Pin 2 SA1 Missed				All Faults Detected			
A	B	C	D	A	B	C	D
1	0	1	1	0	1	0	0
0	0	0	1	1	1	1	1
0	1	0	0	0	0	0	1

FIGURE 11-22   All paths to the output are not sensitized.

11.9 Trade-offs on Selecting Faults · 203

faults only. If path-sensitizing techniques are used, patterns will also pick up equivalent and simultaneously detected pin faults. Thus, an efficient procedure is to declare I/O faults and prepare a pattern set that detects a high percentage of them. Node faults can be added; this will cause a larger set of faults to be evaluated, lowering the percentage detected. Patterns can now be added to increase detection to an acceptable level. The process can be repeated with pin faults, and a high-quality test can be achieved.

Short-circuited pins are another real-world fault. They can be simulated, but as in the case of internal faults, the total number for all combinations of short circuits becomes very large and increases computation. Short circuits actually manifest themselves as multiple node faults, which are detected either singly or simultaneously. In the latter case, the signature will not match; however, the fault is detected and can be isolated using the known good node states and a computer-guided probing algorithm. A poorly selected pattern set may not detect all short circuits if nodes are moved simultaneously. In Figure 11-23, pins D and E are short-circuited to each other but not to ground or to V_{cc}. The outputs are not treated individually when programmed by the bad pattern set from A to B. Nodes D and E thus do not move to opposite states, and a short circuit would not be detected. A rigorous approach to pattern selection by path sensitization will detect all node faults, pin faults, and short circuits. Applying patterns with care to

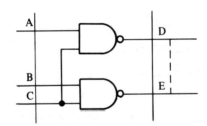

Bad Patterns		Good Patterns	
ABC	DE	ABC	DE
001	11	111	00
111	00	011	10
000	11	101	01
110	11	110	11

FIGURE 11-23 Short-circuit fault from D to E.

204 · Modeling and Simulating Faults

avoid ØØ to 11 transitions on circuits where short circuits are likely, probably costs less than simulating all possible short-circuit faults, when compute times and additional pattern generation times are taken into account.

11.10 Possible Fault Detection

With reference to the circuit shown in Figure 11-24, if a failure G2 PIN3 SA1 exists, the circuit will not initialize. Thus, pattern 2 only detects the fault if the original states of C and D were 1Ø; otherwise, the test passes. This condition is called a *possible detect*. The fault will be detected 50% of the time if there is no preferred state of the flip-flop. A possible detect cannot cause elimination of a fault machine; however, if the flip-flop is randomly moved by various patterns and is possibly detected many times according to the simulation results, it may be declared detected. Practically speaking, 16 possible detects produce a very low probability of missing the fault.

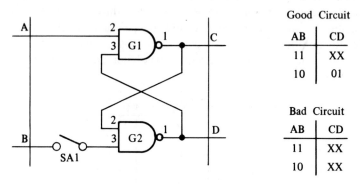

FIGURE 11-24   A possible detect situation.

If the state of a circuit output is indeterminate, detection of a fault via simulation is meaningless; however, the hardware will recognize that the state of the circuit is either Ø or 1 by comparing the actual output to the expected output. Thus, if the C-output of a flip-flop is X, and a test fails with an expected value of Ø, the state of the signal is 1. This fact can be anticipated by the simulator if the hardware stops on the first fail. The table shown in Figure 11-25 gives the results for a set of patterns if we assume initialization in the event that the test passes. If test 2 passes, we can assume that C and D are Ø1. Since test 3 does not change the output, the Ø1 on test 2 can be assumed to

Pattern	AB	CD
1	11	XX
2	10	XX
3	11	XX
4	01	10
5	11	10
6	10	01

Faulty Circuit

Pattern	AB	CD
1	11	XX
2	10	01
3	11	01
4	01	10
5	11	10
6	10	01

Good Circuit

Pattern	AB	CD
1	11	XX
2	10	XX
3	11	01
4	01	10
5	11	10
6	10	01

Faulty circuit, stop on first fail

FIGURE 11-25 Patterns resulting if initialization is assumed.

be the expected good output. If a fault signature contains only one output pin that fails, the failed pin is opposite the good machine state. If a possible detect occurs, X appears in the signature.

```
- - - - - 1 - - -      Absolute detect
- - - - - X - - -      Possible detect
```

In this case, X does not signify an indeterminate state; if the known good response is Ø, the fault signature is obviously 1. The X signifies that passage of the test does not mean that the fault has been detected. In the event that multiple pins have failed:

```
1 0 1 1 0 0 1      Good pattern
- - X X - - -      Fault signature
- - Ø Ø - - -
- - Ø 1 - - -      Possible signature
- - 1 Ø - - -        combinations
```

The use of X in the fault signature saves storing all of the possible signatures and is thus a form of shorthand. The guided probe algorithm uses the known states of the board to determine whether actual observed states constitute a fault signature match.

11.11 Testing Memory Elements

As noted previously, faults are not inserted internal to memory primitives. If we are using a ROM or RAM, we do not simulate the failure of an internal data word, nor do we simulate all failed address words. Level 1 faults are inserted on connector pins so that faults on all device inputs and outputs are presented. It is incumbent upon the test programmer who wants to, for example, test the contents of a ROM during a PC board test, to specifically clock out all data locations. While it is necessary to have correctly modeled ROM's contents, it is not necessary to specifically examine the data coming out. The simulator will process the data and compute the correct known good output response. In the case of a ROM, the test engineer must clock a checkerboard or other test sequence although the foregoing test is only capable of determining and testing the logical functioning of a memory device in a PC environment. This should not be confused with memory testing, where sophisticated test equipment is used to check parameter and dynamic response of memory devices.

12 •

TEST PATTERN GENERATION

12.1 Path Sensitization

Software pattern generators have been developed sufficiently to automatically provide 100% tests for combinatorial circuits. The approach is to begin at each output pin and systematically trace "sensitive" paths to the inputs, so that toggling input pins will toggle the output. All faults along the path will then be detected. When all output–input paths have been sensitized and toggled, all faults will be detected. A NAND-gate is "sensitive" when all except one input are set to 1's. The remaining input then controls the output. (In the case of an OR gate, all except 1 input is set to a Ø.) A systematic test of the circuit shown in Figure 12-1 is generated as depicted in Figure 12-2.

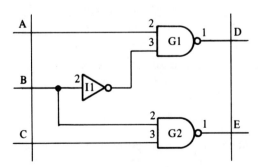

FIGURE 12-1 Circuit for sensitive-path analysis.

Path/States	Pattern	
	ABC	DE
1. D-A/1, E-B/0, D-B/0	101	01
2. D-A/0	001	11
3. E-C/1, E-B/1	011	10
4. D-B/1	111	10
5. E-C/0	110	11

FIGURE 12-2 Test pattern for configuration of Figure 12-1.

We first sensitize the path from output D to input A. The notation in the pattern indicates the output–input path followed by a slash and the state of the driver node. There are four paths. When we obtain a sensitive path, we toggle the driver node to the opposite state. Systematically working through all paths, D-A, D-B, E-B, and E-C, enables us to test the circuit in five patterns. Three paths are set up by the first pattern, but toggling D-A to Ø initializes them, and they must be reestablished. Figure 12-3 shows an example of a reconvergent path

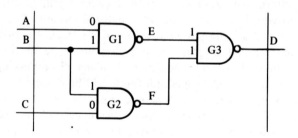

A	B	C	D	Path
0	1	0	0	AED/0
1	1	0	1	AED/1
1	0	0	0	BED/0
0	0	1	0	BFD/0
0	1	1	1	BFD/1
0	1	0	0	CFD/0

FIGURE 12-3 Example of reconvergent path.

(BED and BFD). These occur in redundant logic and produce conflicts. In this case, a path cannot be sensitized. In the case of Figure 12-3, the logic function is nonredundant, and sensitization is possible. On the other hand, in Figure 12-4, it is not. Path CFD cannot be sensitized, because setting A to 1 causes E to go to zero. The logical function is $C = A$.

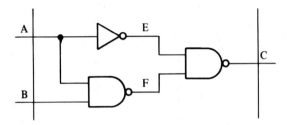

FIGURE 12-4   Sensitization is not possible in this configuration.

Pattern sensitization is greatly complicated by sequential logic containing memory elements. While the principles can be applied, all outputs must be observed by clocking the desired data to the board edge. This may unsensitize the path, and another path, if available, must be chosen to move the required nodes. The example shown in Figure 12-5 applies sensitization to a JK flip-flop. Note that the clock is used to transfer out J and K faults. Figure 12-6 lists the 100% test results for a JK flip-flop.

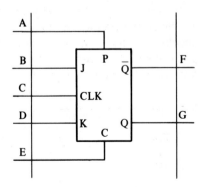

FIGURE 12-5   Sensitization applied to a JK flip-flop.

210 · Test Pattern Generation

P C	J K	CL	Q	Q̄	
1 1	1 1	1	X	X	
0 1	1 1	1	0	1	PQ/0
1 1	1 1	0	1	0	PQ/1
1 1	0 1	1	1	0	QJ/0
1 1	0 1	0	0	1	QJ/0
	1	1	0	1	QJ/1
		0	1	0	QJ/1
	1 0	1	1	0	QK/1
	1 0	0	1	0	QK/1
	1 1	1	1	0	QK/0
		0	0	1	QK/0
1 1	1 1	1	0	1	QC/0
0		1	0	1	QC/0
1 1	1 1	0	1	0	QC/1

100% Test of JK Flip-Flop

FIGURE 12-6 Results of 100% test for a JK flip-flop.

12.2 Initialization

The simulator gives all memory elements X outputs until they are set to known states. It is necessary to initialize a node prior to absolutely detecting its faults. If the PC board uses a capacitive pull-up (Figure 12-7) to initialize a node or ties presets and clears to V_{CC}, initialization can be difficult. There are two requirements: initializing hardware and initializing software. Hardware is automatically initialized prior to the first software pattern. Since the preset pin cannot be controlled by the DTU, there is no means by which to change the state of the model. One way, assuming the connection can be made, is to bring a DTU pin to a test point and use it to pull down the preset. With reference to

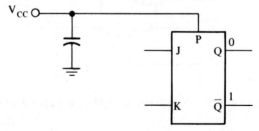

FIGURE 12-7 An example of initialization difficulty.

Figure 12-8, TP1 is declared as a pin on a fictitious connector C*. TP1 on the hardware is connected to a DPU pin, which is brought though a 1-Ø-1 sequence to initialize the device.

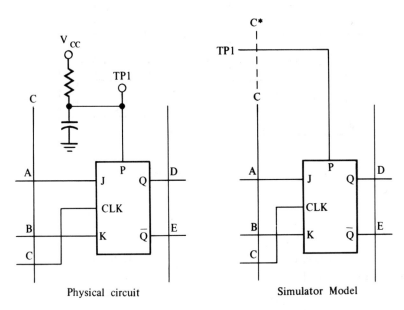

FIGURE 12-8 DTU pin used to pull down a preset.

If it is not possible to connect the preset pin to a DTU pin, it is possible to initialize the simulator prior to applying the first pattern to the known state set by the hardware power-up circuit. When this is done, nodes in the model are changed from their initial X state to the desired 1 or Ø. It is important when using this technique to remove all X's internal to the model so that they are not clocked out later. If a circuit with a feedback loop is not initialized, it is not always possible to eliminate X's if presets are not available. This situation is illustrated in Figure 12-9.

The circuit can come up in three states of Q1 and Q2: 1Ø, ØØ, and Ø1. The 11 state cannot exist because the clear line resets FF1. The set of patterns given in Figure 12-10 initializes any of the possible hardware states to a final ØØ state, yet when the X-state is used for the computation by the simulator, it propagates through the feedback loop, latching it to X. To initialize the circuit, the following procedure is employed:

1. Attempt if possible to block the output of Q2 by applying X's

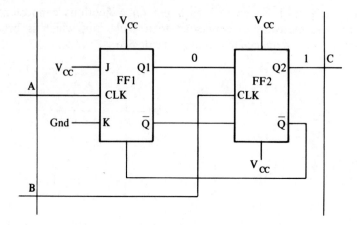

FIGURE 12-9   Feedback loop must be initialized.

Hardware 1			
CL1	CL2	Q1	Q2
1	1	1	0
1	0	0	1
0	1	0	1
1	0	0	0
1	1	0	0
1	0	0	0

Hardware 2			
CL1	CL2	Q1	Q2
1	1	0	0
1	0	0	0
0	1	1	0
1	0	0	1
1	1	0	1
1	0	0	0

Hardware 3			
CL1	CL2	Q1	Q2
1	1	0	1
1	0	0	0
0	1	1	0
1	0	0	1
1	1	0	1
1	0	0	0

Software			
CL1	CL2	Q1	Q2
1	1	X	X
1	0	X	X
0	1	X	X
1	0	X	X
1	1	X	X
1	0	X	X

FIGURE 12-10   States that can be assumed by the hardware.

to any gates that may be in series with the circuit output. (See Figure 12-11.) This prevents the simulator from initializing states by computing a different sequence out of FF2 than that actually computed by the hardware.

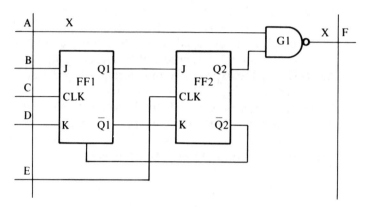

FIGURE 12-11   Output of Q2 is to be blocked.

2. The software model is now initialized to one of the states that can be assumed by the hardware, shown in Figure 12-10. If we force the flip-flop to Q1 = 1, and Q2 = Ø, respectively, the sequence in the hardware table 1 will occur when we apply the initialization patterns. Note that all internal flip-flop states must be initialized. If we do not initialize Q1, it is possible that an X could later propagate out.

Figure 12-12 shows that case where we should initialize on both sides of a delay. If we set signals G3-1 to a 1 and G4-1

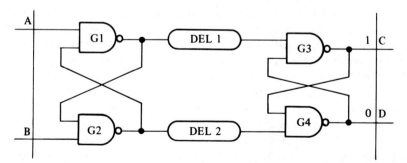

FIGURE 12-12   Both sides of the delay should be initialized.

214 · Test Pattern Generation

to a Ø, the output will be 1Ø, and as soon as computation occurs, the delay circuits will be set to X, thereby forcing an X out of the circuit. Thus, G1-1 and G2-1 must also be initialized to 1 and Ø, respectively.

3. The next step is to apply the pattern sequence that we know initializes the hardware. The model, assuming the hardware 1 initialization state of 1Ø, will clock out the sequence indicated. If the physical circuit is initialized differently, as for example ØØ, the hardware 2 sequence will be clocked out. Since these are different, the expected results during a board test are not useful for detecting faults. We therefore turn off board machines during simulation of these patterns. Note that any memory elements that store the pattern sequence may save this erroneous data. They should be set to states known by the models prior to turning on fault machines.

12.3 Fault Location and Simulation Results

As pointed out previously, the simulator computes the output of both good and bad machines for all input patterns. We have thus far been concerned with obtaining an accurate representation of good and faulty physical circuit behavior and with detecting any faults present. The result of using the simulator to do this is four data files. The *pattern file* is built from those patterns selected by the test engineer of a software pattern generator that contribute to fault detection. The test engineer will work with the simulator until a good test, say 95% detection, has been accomplished. As the contents of the pattern file are simulated, the state of every internal node in the circuit is calculated for both good and faulty machines.

The good machine states are collected in a *node state file* and serve as the reference for the good board response. The node state file is the primary information that enables fault location. The outputs of faulty machines that detect faults are collected in the *fault signature file*. Fault signatures provide a fast, although not always accurate, indication of fault location. Finally, the simulators input processor creates the *topology file*. This file contains the names of each device and pin and the interconnection information among them. Thus, in Figure 12-13, if we want to know which device drives U4 PIN6, the topology file would indicate U3 PIN1. The topology file contains only level Ø information.

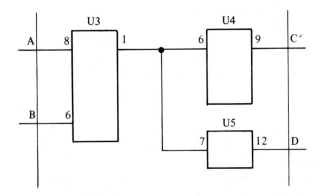

FIGURE 12-13 Configuration for topology file example.

12.4 Post Processors

At the end of simulation, the file data is in a format that is convenient for use by the simulator. A post processor is a file-formatting program that changes the data into a form that can be used by the test system hardware. The pattern file will be organized to consist of both input and computed good-matching output patterns and may be formatted into a source-language program that is input to a language processor which is part of the test system hardware. The remaining files will be compressed as much as possible and stored on a medium such as a disc cartridge.

The post processor combines all files into a *fault dictionary*. This is a document that describes everything occurring on the PC board at the moment a given pattern is applied. The input pattern, expected output pattern, internal node states, all fault signatures, and faults that they detect are provided. The fault dictionary can be used to fault-isolate a circuit manually; however, this is a slow process. If computer-guided fault location algorithms are available, the fault dictionary is used chiefly to verify that the physical circuit performs as the simulator expects.

12.5 Interface Adapters

The interface adapter makes the electrical connection between the DTU pins and the input and output pins of the printed-circuit card, and between the supply pins of the board under test, V_{cc} and ground, and the UUT power supplies in the system. It consists of a printed-circuit inter-

216 · Test Pattern Generation

face card that is inserted into a mechanical fixture between the DTU and the UUT. If V_{cc} and ground pins are identical for all cards, a single fixture can accommodate all UUT's; otherwise, several fixtures or one fixture installed with switching would be used. If the UUT has open collector outputs, the fixture might be used to mount pull-up resistors.

12.6 Test System Operating Software

The automatic test system operates under an operating hierarchy that receives operator commands, executes a driver that runs the DTU, and executes as a fault-isolation software package that operates when a test fails. Figure 12-14 shows a block diagram of a particular system. The normal software execution sequence is shown in Figure 12-15. The fault-isolation software is activated when a test fails. These routines analyze signature data to determine the location of potential failures and then to initiate probing to determine failed nodes.

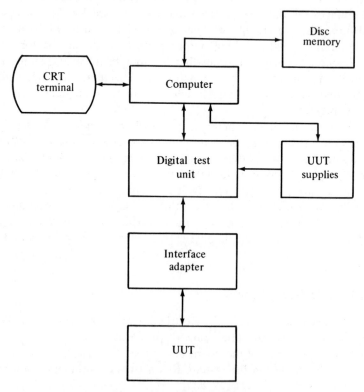

FIGURE 12-14 Digital test system.

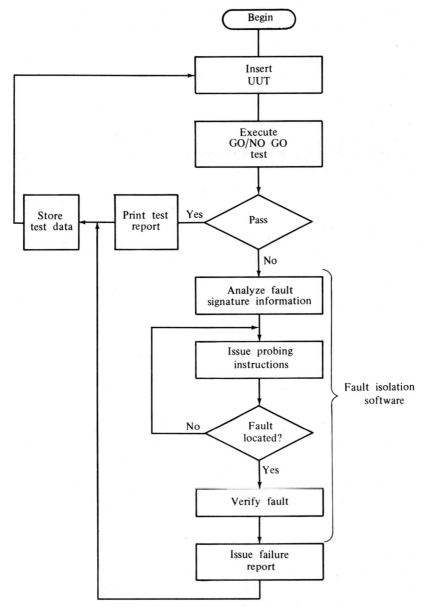

FIGURE 12-15 Software execution sequence.

12.7 Fault Location

If test patterns do not detect a fault, the system will signal "Pass" and fault location does not take place. For this reason, the bulk of this discussion is concerned with the preparation of the pattern set. Once a fault has been located using a pattern set prepared with a logic simulator, we have all the data needed to perform fault location. A failure is detected when one or more pins are in opposite states from those expected. These failed pins are the starting point for fault location. A "path" of failed states from the failed device to the board edge exists at time of failure. This situation is illustrated in Figure 12-15. If the failed output is Ø, each input to gate G3 is probed, and it is found that PIN2 is failed SA1. Probing the input node to G2 indicates PIN2 SAØ. Probing inputs to G1 indicates that both pass. In turn, the failed node is G1-1. The process is more complex, but just as accurate, if sequential logic is being tested (Figure 12-16).

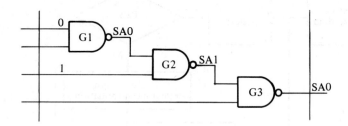

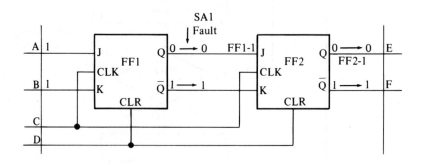

| | Good Circuit | | | | | Faulty Circuit | | | | |
Pattern	A B C D			E	FF1-1	A B C D			E	FF1-1
1	1 1 0 0			0	0	1 1 0 0			0	1
2	1 1 0 1			0	0	1 1 0 1			0	1
3	1 1 1 1			0	0	1 1 1 1			0	1
4	1 1 0 1			1	1	1 1 0 1			0	1

FIGURE 12-16 Fault isolation of sequential logic.

In pattern 1, clock and clear are set to Ø, forcing signals FF1-1 and FF2-1 to ØØ. Pattern 2 raises the clear line; pattern 3 raises the clock line and thereby transfers the state of Q and $\overline{Q}$ of FF1 into the J and K inputs of FF2. The correct response would be Ø1, and when the clock is lowered by pattern 4, the Q-$\overline{Q}$ outputs of pattern 4 are unchanged. If a SA1 failure occurs in FF1, the pattern 11 is transferred into FF2, and a 1Ø appears at the output when the clock drops. The fault, present in the circuit through patterns 1, 2, and 3, is clocked out in pattern 4. In sequential circuits, failures are propagated to the edge of the board, usually via clock pulses, and thus are detected on earlier patterns if internal nodes are probed. It is therefore insufficient simply to measure the state of the board at the failed pattern; all patterns are recycled at each probing attempt, and comparisons are made for each.

12.8 Feedback Loops

In a feedback loop, many devices, only one of which is bad, can be latched into a circular failure path. The feedback loop in Figure 12-17 illustrates this situation. The failed node is G1-1 SAØ. Following the algorithm of probing each driving device until a good node is found, we would probe in the following sequence:

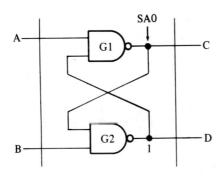

Correct Response				Failed Response			
A	B	C	D	A	B	C	D
1	1	0	1	1	1	0	1
0	1	1	0	0	1	0	1

FIGURE 12-17 Configuration with a feedback loop.

220 · Test Pattern Generation

PIN	STATUS
Pin C (U1-1)	SAØ
U1-1	Good
U2-1	SA1
U2-3	Good
U1-1	SAØ

Upon detecting the repeat probe, U1-1, we know that we have a feedback loop consisting of failed driver nodes U1-1 and U2-1. It is not possible to determine which of these nodes has failed, yet we do know that both were frozen for both patterns, and we can expect a "stuck at" fault. We could encounter a situation in which a short circuit between pins causes a failing feedback loop that moves as patterns are cycled. After probing the entire feedback path, it can be identified in the test report, enabling the operator to remove all suspect faults. An interesting effect can occur in the event that the failure is both output nodes (Figure 12-18). The short circuit forces G1 and G2 into the active region, producing a "stuck-in-the-middle" failure. This failure can propagate to the next IC but will not do so consistently. The logic probe hardware senses this condition as a "misprobe" since the voltage read is in the quiescent bias zone of the input amplifier.

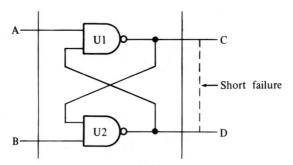

FIGURE 12-18   An example of both output nodes short-circuited.

12.9   A Fault Isolation Probing Algorithm

In Figure 12-19, we have departed from usual notation and have numbered gates on two devices, U1 and U2, with numbers as they appear on a standard dual in-line package. The pattern shown in Figure 12-20 detects the following faults, all having the signature 11. Figure 12-21 shows the probing algorithm. The signature tells us that the first possible failed driver is not U2 PIN3 or U2 PIN4, but U2 PIN10. U2

12.9 A Fault Isolation Probing Algorithm · 221

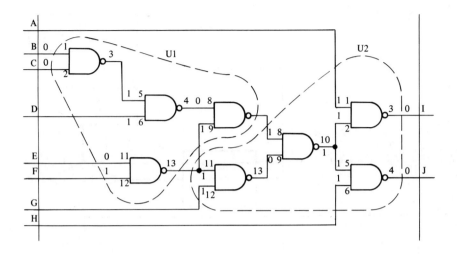

A B C D E F G H	I J
1 0 0 1 0 1 1 1	0 0

FIGURE 12-19 Example of probing sequence.

$$
\begin{array}{l}
\text{Equivalent} \left\{
\begin{array}{ll}
\text{U2 - 10} & \text{SA0} \\
\text{U2 Pin 9} & \text{SA1} \\
\text{U2 Pin 13} & \text{SA1} \\
\text{U2 Pin 11} & \text{SA0} \\
\text{U2 Pin 12} & \text{SA0} \\
\text{C - 6} & \text{SA0}
\end{array}
\right. \\[2em]
\phantom{\text{Equivalent} \{} \text{U1 - 13} \quad\quad \text{SA0} \longleftarrow \text{Actual Failure} \\[1em]
\text{Equivalent} \left\{
\begin{array}{ll}
\text{U1 Pin 11} & \text{SA1} \\
\text{C - E} & \text{SA1}
\end{array}
\right.
\end{array}
$$

FIGURE 12-20 Fault isolation probing pattern.

PIN10 is the first pin probed. It fails, and the two previous drivers, U1 PIN10 and U2 PIN13 are probed. U1 PIN10 passes, because the gate is not sensitive. U1 PIN13 fails, and C pins E and F pass. We thus suspect a failure of U13 Ø and a failed U1 device. We reprobe U1

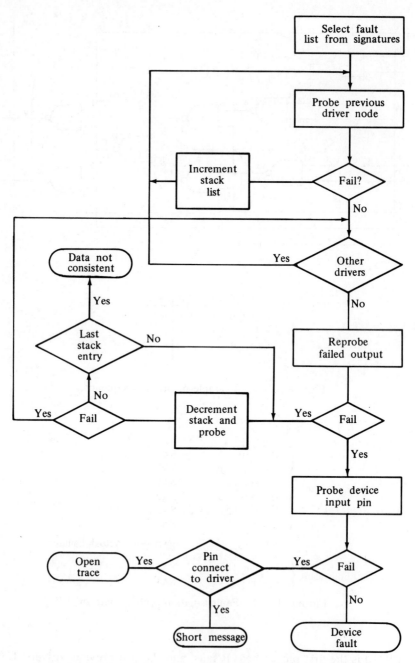

FIGURE 12-21 Probing algorithm for fault isolation procedure.

PIN13 to confirm the failure. Should it not be repeated, we would decrement the stack of probed nodes and reprobe previous ones. This is a mechanism for eliminating probing errors. The failure path is retraced until consistent data are found. When a failure is confirmed by a second probing, device input pins are probed to ensure that an open trace between the previous driver and the input is not present. All remaining inputs are probed to determine if other failure paths or short circuits exist.

12.10 Parallel Drivers

The circuit of Figure 12-22 shows a case in which a pin fault in one of several driver devices fails, but probing driving nodes fails to locate the correct fault. The probe sequence would be as follows:

U1 PIN1	(Fail)
C PINA	(Pass)
C PIND	(Pass)

The solution is to recognize that pin D fans out to various device inputs and that each one of these must be probed.

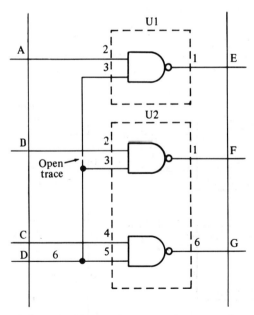

FIGURE 12-22 Probe sequencing example.

12 . 11 Internal Device Feedback

Certain models may contain paths that propagate faults to more than one output pin. In Figure 12-23, the circuit is assumed initially not to include A1 and A2. The output signature of an unbuffered model has two failed paths, because the faulty node drives G2, causing a multiple fault effect on C and D that will not have a signature match. If the guided probing algorithm picks the failure path to pin D, the failure will be incorrectly indicated as G2 PIN1 SA1. A converse effect occurs if the model is buffered and the device is not. The actual device will have two failure paths and will produce an unmodeled signature. Figure 12-24 summarizes each possibility.

Expected output	Unbuffered signature	Buffered signature
1	0	0
0	1	0

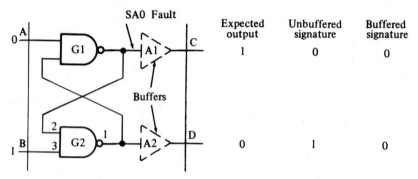

FIGURE 12-23 Faults may be propagated to more than one output.

Good response	Signature
1	0
0	1
1	0
0	0

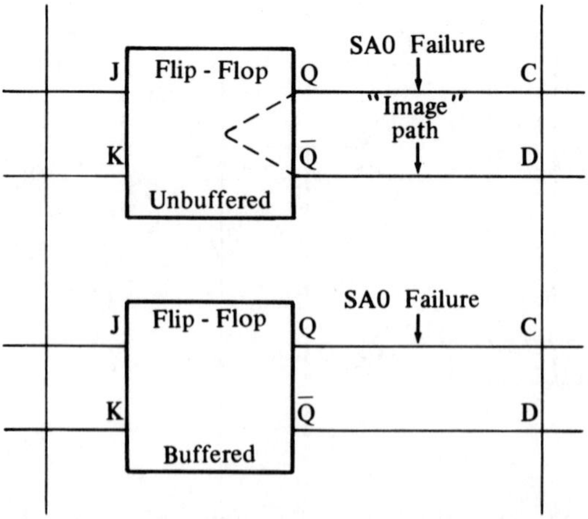

FIGURE 12-24 Example of an unmodeled signature situation.

Buffered model/ buffered real:	Single failure path. The correct fault is detected.
Unbuffered model/ Unbuffered real:	Signature match indicates two paths; the state follows one of the paths. If the "image" path is selected, one wrong mode may be displayed as failed.
Unbuffered model/ Buffered real:	A single failure path to the edge. There is no signature match, since the image path does not actually occur the probing algorithm goes to the correct fault.
Buffered model/ Unbuffered real:	There is no signature match, since the real device has an "image" path. The image path, if selected, will lead to display of wrong node, as failing.

FIGURE 12–24   *Continued*

12.12   Pulse Response

As noted previously, each test is static, in the same sense that the DTU reads the signal output after all nodes have settled. In the event that a circuit's correct operation is dependent on the existence of a pulse, a static test will not adequately find the failure. In Figure 12-25, the failure of a pulse may cause Q to take an incorrect state. A probe sequence that did not account for pulses would show that signals C-A, C-D, and DEL1 are statically correct and erroneously indicate that FF1-1 failed. The probe is therefore designed to read static levels,

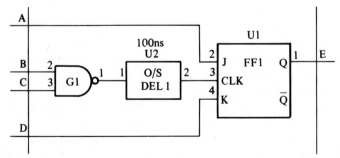

FIGURE 12-25   Arrangement for probing double transitions.

226 · Test Pattern Generation

transitions, and double transitions, or pulses. This enables the algorithm to present data on a specific node in the format shown in Figure 12-25.

The presentation in Figure 12-26 indicates that during a board test, changes occur from pattern 1 to pattern 40. The node toggles on patterns 2, 6, 12, 36, 37, and 40. Pulses occur on patterns 28 and 37. If a circuit does not depend upon pulses for correct operation, we want to ignore them. During testing, a trace adjacent to a fast-changing signal may acquire a "glitch," or pulse. If the hazard does not affect circuit performance, it can be ignored. This is the normal mode of the guided probe software. When we want to consider pulses, as in the circuit of Figure 12-25, the node state file can be augmented with pulse information. For example, this can be done in tests 28 and 37. Absence of a pulse will then correctly identify the one-shot multivibrator as the failed component.

U2 Pin 2

Test	State	Pulse			
1	0	0			
2	1	0			
6	0	0			
12	1	0			
28	1	1	28	1	0
36	0	0			
37	1	1			
40	0	0			

Good Node   Failed Node

FIGURE 12-26 Node activity description.

12.13 Operator-Directed Probing

Guided probing algorithms are not perfect. Although in practice they can adequately locate most faults, certain conditions may require the test operator, using a schematic, to probe any circuit node. In this instance, the short form test report for a node is displayed to the operator as in Figure 12-26. Operator-directed probing is particularly valuable in instances where unexpected pulses affect circuit operation. In the case of Figure 12-25, a fault would be detected and located and the flip-flop replaced; however, on retest, the failure would reappear.

Operator-directed probing would indicate that pulse response must be considered on U2 PIN2.

12.14 Validation

Hardware validation consists of probing each node to ensure that both static and pulse response are correct. Pulse response discrepancies have been described previously; the typical cause of a static response error is a modeling error, or schematic error. Since considerable time and financial investment has been made in the test program prior to hardware validation, verification of models using procedures explained in the foregoing discussions are extremely important.

APPENDIX I •

TROUBLESHOOTING THE INTEL 4004 MICROPROCESSOR SYSTEMS

This application note, prepared by the Hewlett-Packard Co., demonstrates real-time analysis of program flow and triggering on specific data events, as well as paging techniques for the Intel 4004 microprocessor systems. This microprocessor operates from +5-volt and −10-volt power supplies. It features a 4-bit parallel CPU with 46 instructions. The 4004 can directly address 4k 8-bit instruction words of program memory and 5120 bits of data storage RAM. Up to 16 4-bit input ports and 16 4-bit output ports can also be directly addressed. Sixteen index registers are provided internal to the microprocessor for temporary data storage. The 4004 microprocessor operates at clock rates to approximately 750 kHz. A diagram of its pin assignments is shown in Figure I-1.

Note that the D0-D3 pins connect to a bidirectional data bus that handles all address and data communication between the microprocessor and RAM and ROM chips. The Ø1-Ø2 pins are driven by nonoverlapping clock signals that determine the microprocessor timing. Next, the Sync pin is for a synchronization signal that indicates the beginning of an instruction cycle to the ROM and RAM chips. The Reset pin is for application of a "1" level to clear all flag and status flip-flops and to force the program counter to 0. Reset must be applied for 64 clock cycles (8 machine cycles) to completely clear all address and index registers. Observe that the Test pin provides examination of the logic

230 · Appendix I

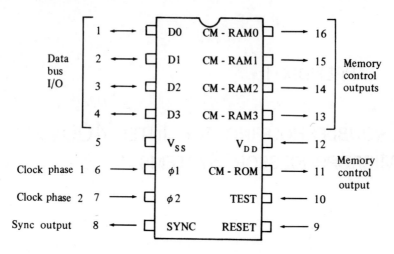

FIGURE I-1 Pin assignments for the Intel 4004 microprocessor.

state of "test" (input) with JCN instruction. The CM-ROM pin connects to a line that enables a ROM bank and I/O devices that are connected to the CM-ROM line. Also, the CM-RAM0 through CM-RAM3 pins connect to lines that function as bank select signals for the RAM chips in the system.

Consider next the probe connections that are used with the HP-1600A and 1607A logic state analyzers. The 4004 microprocessor does provide a unique clock for the logic state analyzer at the proper time (end of AS state) in the instruction cycle. The CM-ROM line is always true at A3 and can be used as a clock signal. However, CM-ROM also occurs at states M2 or X2 during the execution of some instructions. This would result in invalid data being displayed by the analyzer. By constructing the circuit shown in Figure I-2, the troubleshooter can ensure a correct state display. If the portion of the program that is to be examined is completely contained on one ROM, the chip-select line (CS) for the ROM can be used as a clock.

A system that will not "come up" can frequently be debugged by monitoring address flow alone. The 4004 CPU chip has a 4-bit data bus, on which the 12-bit address is multiplexed during A1, A2, and A3 states of the 4004 machine cycle. In order to view the demultiplexed 12-bit address on a 1600A, the 4004 system must use the 4008/4009 Standard Memory and I/O Interface Set, the 4289 Standard Memory Interface, or similar logic circuits that provide a demultiplexed address bus. If a particular system uses memory chips that internally decode

Appendix I · 231

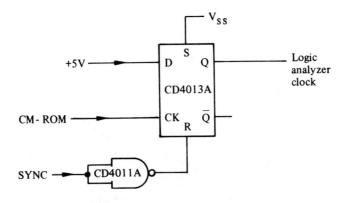

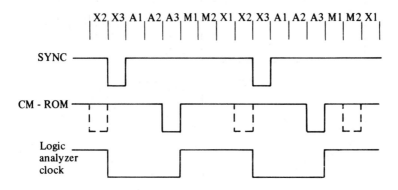

FIGURE I-2 Circuit for deriving a clock signal for the HP 1600A from 4004 Sync and CM-ROM signals.

the multiplexed address, such as the 4001 ROM, the troubleshooter should monitor the microprocessor data bus as described subsequently. Probe connections as shown in Figure I-3 provide a display of the activity on the address lines.

Next, consider the control settings. With the power switch for the logic state analyzer turned on, the troubleshooter sets its controls as follows: Display Mode, Table A; Sample Mode, SGL. Note that SGL is selected for viewing single-shot events. Press RESET to start the system. The first time that the system passes through the trigger point, the display will be generated and stored. For programs that are looping or cycling through the selected address, select REPEAT sample mode. The Trigger Mode is set as follows: NORM/ARM, NORM; LOCAL/

232 · Appendix I

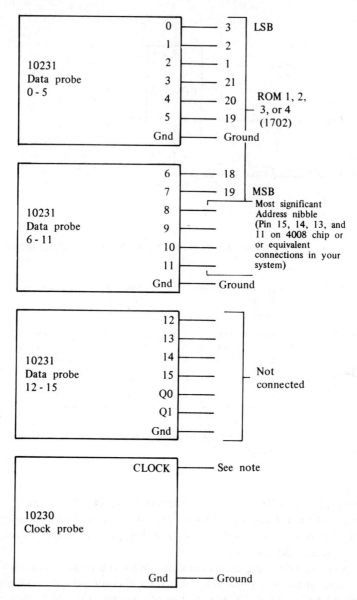

FIGURE I-3 Probe connections for display of activity on the address lines.

BUS, LOCAL; OFF/WORD, WORD. Threshold is set to VAR and adjusted to 3.7 volts. Note that when troubleshooting TTL compatible systems, the threshold is set to TTL. The other controls are set as follows: Logic, POS; Clock, ⌐; all other pushbuttons, Out position; Display Time, ccw; qualifiers Q1 and Q0, set to OFF; Trigger Word Switches, set to address at which it is desired to trigger; Column Blanking, after display is on screen, adjust blanking to display 12 columns of data.

Next, consider interpretation of the display. In this example, a segment of a chip tester program for Quad NAND gates is examined. Proper operation is confirmed by a comparison between real-time state analysis as shown in Figure I-4(a) and the 4004 cross assembler program listing in Figure I-4(b). The chip tester routine performs the following events: (1) sets up bit patterns in the accumulator; (2) outputs the accumulator contents to the NAND gates (connected to I/O Port 1); (3) reads the gate outputs; and (4) tests on the gate outputs and indicates whether the chip is good or bad.

Observe the program listing in Figure I-4(b). The instruction located in addresses 030 through 033 load the bit pattern 0010 into the accumulator. The next instruction (WRR), in location 034, writes the accumulator contents to output Port 1. The next two instructions (address locations 035 and 036) read the gate outputs present at input Port 1 into the accumulator and complement the accumulator. Examination of lines 1 through 7 of the state display shown in Figure I-4(a) shows that these instructions have been executed in the proper sequence.

The instruction starting at address 037 is a conditional jump that is a two-word instruction (lines 8 and 9 of the state display). If the chip passes the test (accumulator contains all zeros), the program continues the test routine. If the chip fails the test, the program jumps to an output routine. Examination of line 10 of the state display, address 039, reveals that the chip passes the test. The program then outputs another bit pattern (1111) to the chip under test and reads the input port. This is shown by lines 10 through 13 of the state display. Lines 14 and 15 of the state display are the addresses of the two words of another JCN instruction. Line 16 of the state display is the address 043, showing that the chip failed the last test, causing the program to jump to the output routine.

Next, consider the map. If a tabular display is not presented in the preceding operation, it means that the system did not access the selected address and the No Trigger light will be on. To find where the system is residing in the program, switch to "map" (see Figure I-5). Using the trigger word switches results in moving the cursor (circle in lower right

234 · Appendix I

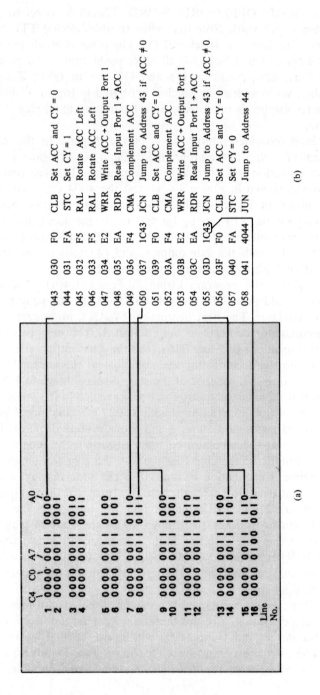

Line No.				
043	030	F0	CLB	Set ACC and CY = 0
044	031	FA	STC	Set CY = 1
045	032	F5	RAL	Rotate ACC Left
046	033	F5	RAL	Rotate ACC Left
047	034	E2	WRR	Write ACC → Output Port 1
048	035	EA	RDR	Read Input Port 1 → ACC
049	036	F4	CMA	Complement ACC
050	037	1C43	JCN	Jump to Address 43 if ACC ≠ 0
051	039	F0	CLB	Set ACC and CY = 0
052	03A	F4	CMA	Complement ACC
053	03B	E2	WRR	Write ACC → Output Port 1
054	03C	EA	RDR	Read Input Port 1 → ACC
055	03D	1C43	JCN	Jump to Address 43 if ACC ≠ 0
056	03F	F0	CLB	Set ACC and CY = 0
057	040	FA	STC	Set CY = 0
058	041	4044	JUN	Jump to Address 44

(b)

FIGURE I-4 System response to test routine.

of screen) to encircle one of the dots on the screen. Switch to Expand, and make the final positioning of the cursor—the No Trigger light will now go out, and switching back to Table A displays the 16 addresses around that point.

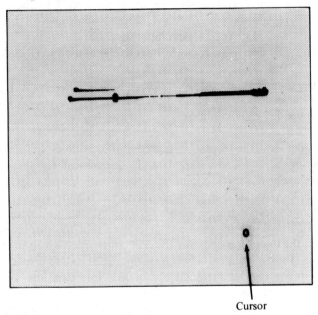

Cursor

FIGURE I-5 Map display shows the entire system activity.

Viewing address, data, and controls are utilized when program deviations are found. The reason may be as simple as a program error or as complicated as a hardware failure on the data bus or command lines. Additional input channels now become very desirable for the troubleshooter. By combining the 1600A and 1607A, the display and trigger capability can be expanded to 32 bits wide, allowing the 12-bit address, 8-bit data word, and up to 12 other active control signals to be viewed simultaneously. The procedure is as follows:

1. Connect the data cable between the rear panel connectors.
2. Connect the trigger bus cable between the front-panel bus connectors.
3. Set the 1600A controls as described previously, with the following exception—set the display mode to Table A&B.
4. Set the 1607A controls as follows: Sample Mode, SINGLE;

Start Display, ON; Trigger Mode—NORM/ARM, NORM; LOCAL/BUS, BUS; OFF/WORD, OFF; all other pushbuttons, Out position; qualifiers Q1 and Q0, OFF.

5. Connect data and clock inputs for the 1607A as follows:
   (a) Connect 1607A data inputs 0 through 7 to the demultiplexed data bus (ROM output) starting with LSB connected to 1607A data input 0;
   (b) Connect 1607A clock input to signal used to clock 1600A; and
   (c) Connect grounds to appropriate points.
6. After a display appears on the screen, set the 1607A blanking to display eight columns.

Consider next the display interpretation of address and data lines. By displaying both address and data, it is practical for the troubleshooter to confirm exact system operation with respect to the test routine. Looking at line 1 in the state display of Figure I-6(a), observe that the data corresponding to address 030 is F0, the 8-bit code for the CLB instruction. Looking at line 2, it is seen that the displayed word agrees with the operation code for the STC instruction given in the program listing. In this manner, subsequent lines of the state display can be examined to show exact program operation. Note that line 14 of the state display corresponds to the first word of the JCN instruction at address 03D. The data in line 15 corresponds to the second word (0100 0011) of the JCN instruction, the address that program control is transferred to if the jump condition is true. Examination of line 16 shows that the program did jump to the specified address.

It is helpful to consider the viewing of the multiplexed data bus. In the preceding examples, the demultiplexed address and data lines have been observed on the ROM address and data lines. However, when a hardware failure occurs, it may be very helpful to the troubleshooter to directly observe activity on the multiplexed microprocessor data bus. In the following example, a description of observed data being demultiplexed into a 12-bit address for driving a ROM is given. Then, an observation of the ROM output being multiplexed back onto the bus will be explained. To start, the troubleshooter sets up the 1600A and 1607A to obtain the display, as follows:

1. Connect the 1600A data, qualifier, and clock inputs in the following way:
   (a) 1600A data inputs 0 through 7 to RD0 through RD7 on the ROM in order;

Appendix I · 237

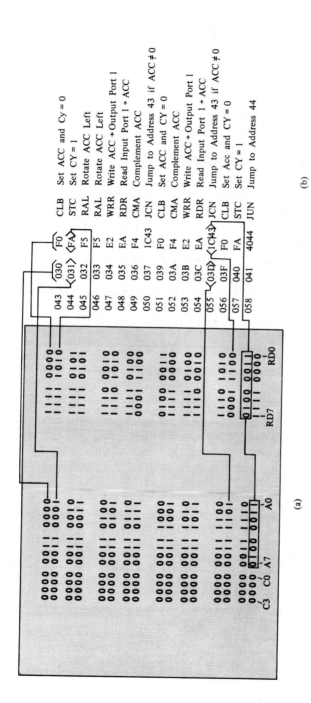

Figure I-6 System response to test routine on address and data lines.

(b) 1600A data inputs 8 through 15 to A0 through A7 in order;
(c) 1600A Q0 input to ROM 0 chip select line (1702A, pin 14). Note that by qualifying on CS and triggering on A0 through A7, the troubleshooter derives a unique trigger that is effectively 12-bits wide with only the eight least-significant digits displayed; and
(d) 1600A clock input same as explained in the beginning of this topic.
2. Connect 1607A data and clock inputs to the microprocessor as follows:
(a) 1607A data inputs 0 through 3 to D0 through D3;
(b) 1607A data input 4 to CM-ROM;
(c) 1607A data input 5 to SYNC; and
(d) 1607A clock input to $\emptyset 2$.
3. Set the 1600A controls the same as previously described, with the following exceptions: Display Mode, Table A+B; End Display, ON; Delay, ON with Delay set to LO; qualifier, TRIG with Q0 set to LO; Column Blanking, ccw.
4. Set 1607A controls as follows: End Display, ON; Delay, ON with Delay set to 8; Logic, NEG. Note that the microprocessor data bus uses negative logic, i.e., the most positive voltage is a logic 0, and the most negative voltage is a logic 1.

Next, consider the display interpretation of the multiplexed data bus. The state display diagram in Figure I-7 shows a comparison of the demultiplexed address and data buses (Table A) with the multiplexed microprocessor bus (Table B). Compare line 8 of Table A (trigger word) with the multiplexed data in Table B. Examination of the Sync line shows that line 6 of Table B corresponds with the instruction cycle state A1. Note that the Sync and CM-ROM pulses are displayed as 1's in the illustration, because the troubleshooter has selected negative logic on the 1607A. Comparison of states A1, A2, and A3 (lines 6, 7, and 8 of the Table B state display) with the trigger words address bits reveal that the demultiplexer has correctly processed the address from the 4004. Similarly, comparison of trigger word data bits RD7 through RD0 with states M1 and M2 (lines 9 and 10 of the Table B display) shows that the multiplexer has correctly processed the ROM data onto the 4004 data bus. Note that the CM-ROM line is true during the M2 state, indicating that the instruction being executed is an I/O instruction.

In conclusion, it is evident from the foregoing examples that efficient troubleshooting of the Intel 4004 microprocessor system is

Appendix I · 239

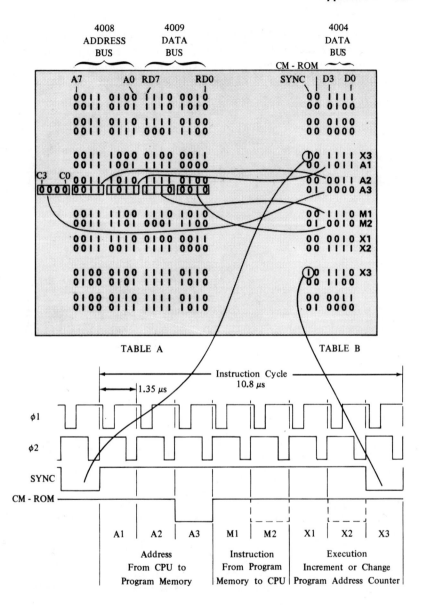

FIGURE I-7 Comparison of 4004 data bus activity with demultiplexed address and data.

expedited by two factors: first, the availability of the program listing as produced by the 4004 cross assembler; and second, the availability of real-time logic state analysis to display actual system operation for rapid error detection and correction.

APPENDIX II •

TROUBLESHOOTING THE MOTOROLA M6800 MICROPROCESSOR SYSTEM

Real-time analysis of the Motorola M6800 microprocessor system in a troubleshooting environment is explained in this Hewlett-Packard application note. It demonstrates the real-time analysis of program flow in positive and negative time, triggering on a specific event and paging techniques. The MC6800 microprocessor, which is the nucleus of the M6800 microprocessor (Figure II-1) family, operates from a single +5-volt supply. The features of this microprocessor include a 16-bit three-state address bus allowing 65,000 addressable bytes of memory and a three-state 8-bit bidirectional data bus. The system, which includes both maskable and nonmaskable interrupts, operates at clock rates to 1 MHz.

Probe connections for the HP 1600A and 1607A logic analyzers are depicted in Figure II-2. A system that will not "come up" can frequently be debugged by monitoring address flow alone. The analyzer probe connections shown in the diagram provide a display of the activity on the address lines. If address bus extenders are utilized, connections are made on the "Extended" side of the bus. Next, consider the control settings. With the power switch of the logic state analyzer turned on, the troubleshooter sets the controls as follows: Table, "A"; Sample Mode, Repet. Note that if the program is not looping or cycling through the selected address, select "Single" and press "Reset" and

242 · Appendix II

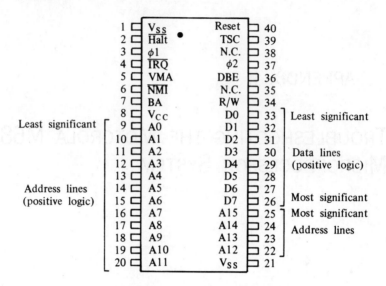

SUMMARY OF CONTROL LINES

	LOW "0" V_{SS}	HIGH "1" $\geq$ 2.4V
$\overline{\text{HALT}}$	All machine activity halted	Machine will fetch and execute instructions
$\overline{\text{IRQ}}$	Interrupt request pending	No interrupt Request
VMA	Address Bus data is INVALID	Address Bus data is VALID
$\overline{\text{NMI}}$	A nonmaskable Interrupt Request is pending	Nonmaskable interrupt is not pending
BA	Address bus is not available	Address bus is available and microprocessor activity has stopped
$\overline{\text{RESET}}$	All microprocessor registers cleared, all machine activity halted	On ⌐ microprocessor executes initial start-up sequence
TSC	Address lines and R/W line are not HI-IMPEDANCE	Address lines and R/W in HI-IMPEDANCE
DBE	Data bus lines are in HI-IMPEDANCE	Data bus drivers are enabled
R/W	MPU "WRITE" operation - data from microprocessor to memory or peripheral	MPU "READ" operation - data from memory or peripheral into microprocessor

FIGURE II-1 Pin assignments for the M6800, with control-line summary.

start the system. The first time that the system passes through the trigger point, the display will be generated and stored.

Continue as follows: Set the Start Display to On; Trigger Mode, "NORM," "LOCAL," "WORD;" for connection directly to the M6800, set Threshold, Var. Adj. to 2.4 volts; Logic, Pos.; for connections to an external bus, set Threshold and Logic to match extender that is used; all other pushbuttons, "out;" Display Time, ccw; Column Blanking, ccw; qualifiers Q0 hi and Q1, off. Note that if it is desired to observe system activity after the completion of the wait for interrupt instruction (WAI = 3E), the Q0 qualifier channel, connected to VMA, should be set to the "off" position, because VMA is low at the completion of this instruction. Trigger word switches are set to the address at which the troubleshooter desires to trigger.

Next, consider the display interpretation. In this example, system response to an interrupt will be explained. Proper operation is confirmed by a comparison between real-time state analysis, shown in Figure II-3(a), and the M6800 cross assembler program listing output, shown in Figure II-3(b). The MC6800 responds to an interrupt according to the following conditions:

1. Complete current instruction.
2. Store microprocessor registers on stack.
3. Load program counter with interrupt service routine vector fetched from memory locations FFF8 (PCH) and FFF9 (PCL).
4. Begin execution of interrupt service routine at vectored location.

Observe the program listing in Figure II-3(b). Note the vector routine beginning at location 274C with the instruction LDA B $1A80. This is a three-byte instruction; the first byte (F6) is the operation code, with the second (1A) and third (80) being a double byte operand, in this case an extended address. Proper operation of the vector fetch is confirmed by observing the address immediately following the intensified trigger word, FFF9, Figure II-3(a), is 274C. This means that the microprocessor fetched 27 from FFF8 and 4C from FFF9.

The LDA B $1A80 may be confirmed by observing the second, third, and fourth lines of the state analysis illustration. Line 2, 274C, is the fetch of the operation code LDA B (F6), and lines 3 and 4 are the fetch of 1A80; 274D contains 1A, and 274E contains 80. The next line, the fifth, shows the address to 1A80, which implies correct execu-

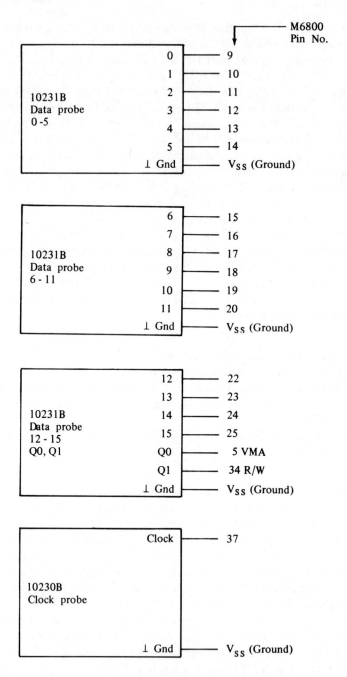

FIGURE II-2 Probe connections for the logic state analyzer.

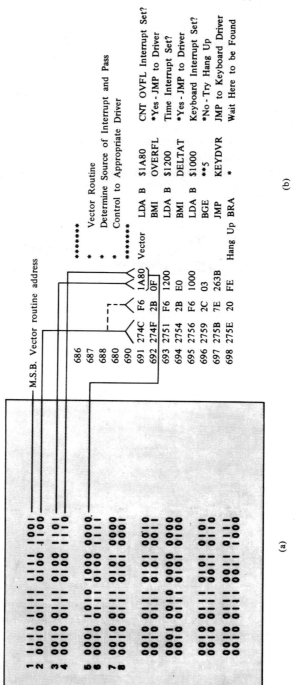

FIGURE II-3 System response to an interrupt.

tion of the instruction in the routine. In a similar fashion, each instruction may be shown to have been properly executed.

Observe the map display in Figure II-4. If a tabular display is not presented in the previous step, it means that the system did not access the selected address, and the No Trigger light will be on. To find where the system is residing in the program, switch to "map." Using the Trigger Word switches, move the cursor (circle in lower right on screen) to encircle one of the dots on the screen. Switch to "expand," and make the final positioning of the cursor. The No Trigger light will now go out, and switching back to Table A displays the 16-bit address around that point.

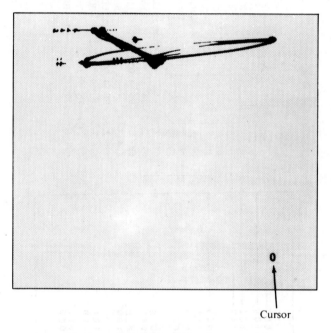

FIGURE II-4 Map display shows entire system activity.

Next, consider the viewing of address and data with the accompanying control settings. When program deviations are found, the reason may be as simple as a program error or as complicated as a hardware failure on the Data Bus, Control Bus, or other command lines. It is desirable for the troubleshooter to have additional input channels at this time. By combining the 1600A and 1607A, the display and trigger capability can be expanded to 32 bits wide, allowing the 16-bit address, 8-bit data bus, and eight other command signals to be viewed simultaneously. Connections are made on the same general principles as exemplified in Appendix I. Interpretation of the display with respect to the address and data buses, read/write, and $\overline{\text{IRQ}}$ lines is made as follows.

Using End Display triggering and digital delay, it is now possible to confirm exact system operation with respect to interrupt response as noted in the previous discussion. With reference to lines 0 through 2 of Figure II-5, observe that the $\overline{\text{IRQ}}$ line is low for three clock periods, which signals the microprocessor to begin executing an interrupt service operation. Confirmation that the microprocessor completed the current instruction is shown in lines 3 and 4 of Figure II-5(a). Similarly, it can be shown that the microprocessor saved the internal registers on the stack by an examination of lines 4 through 11. Line 4 shows the completion of the current instruction at location 2674. Line 5 shows the least significant byte of this address (74) being written into location OFFD while the Read/Write line is low or in the "WRITE" mode. Line 6 shows the analogous operation for the most significant byte of the program counter (26) being written into location OFFC.

The foregoing process is continued until all registers have been stored on the stack, as shown in Figure II-6. It is evident that seven stack locations are required to save the microprocessor status, which may be confirmed by lines 5 through 11 inclusive, in Figure II-5. As discussed previously, the microprocessor then reads memory locations FFF8 and FFF9 to set the interrupt service vector. Observe that the program listing in Figure II-5(b) defines this to be 274C, which means that memory location FFF8 must contain the data 27 and that FFF9 must contain 4C. This is confirmed by lines 12 and 13 of Figure II-5(a). Observe also that the fetch was executed properly, because the next address following FFF9 is 274C (line 14). In addition, the data at location 274C is F6 [Figure II-5(b)], which corresponds with the program listing.

In conclusion, it may be seen from the foregoing examples that efficient troubleshooting of the Motorola M6800 microcomputer system is expedited by two factors: first, the availability of the program listing as produced by M6800 cross assembler, which is the definitive document of program execution; and, second, the availability of real-time logic state analysis to display actual system operation for rapid error detection and correction.

248 · Appendix II

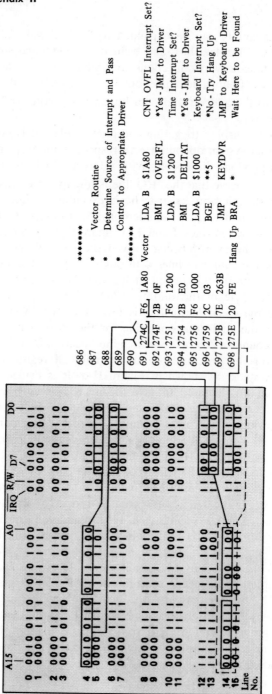

FIGURE II-5 System response to an interrupt on address, data, and control lines.

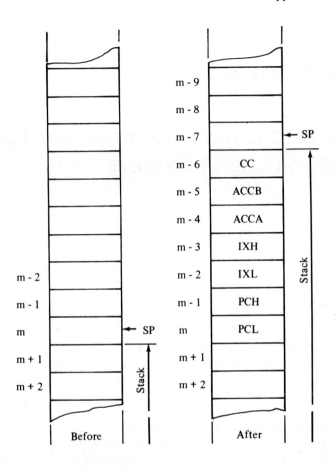

```
   SP   = Stack Pointer
   CC   = Condition Codes (Also called the Processor
          Status Byte)
 ACCB   = Accumulator B
 ACCA   = Accumulator A       Accumulator B
  IXH   = Index Register, Higher Order 8 Bits
  IXL   = Index Register, Lower Order 8 Bits
  PCH   = Program Counter, Higher Order 8 Bits
  PCL   = Program Counter, Lower Order 8 Bits
```

FIGURE II-6 Saving the status of the microprocessor in the stack.

APPENDIX III •

TROUBLESHOOTING THE FAIRCHILD F8 MICROPROCESSOR SYSTEMS

This appendix text is intended to assist the troubleshooter in localizing malfunctions in the Fairchild F8 microprocessor systems; it was originally prepared as an application note by Hewlett-Packard. It demonstrates real-time analysis of actual program sequences, triggering on specific events, selective data qualification, and paging techniques. The microprocessor and read-only memory form a basic two-chip system that will handle most simple tasks. For more complex processing functions, the memory access and interface chips complete a four-chip set. The F8 system is unique in that each chip has its own resident program counter, thus eliminating the need for a dedicated address bus and its associated circuits. Branch commands that require resetting the address counter are multiplexed onto the data bus for implementation. This architecture makes possible a very low-cost system including I/O ports as a standard part of the basic chips. CPU chip pin assignments are shown in Figure III-1. Memory interface chip pin assignments are shown in Figure III-2.

Probe connections to the 1600A and 1607A are made as depicted in Figure III-3. A system that will not "come up" can frequently be debugged by monitoring the address flow alone. With a memory comprising the F8 central process unit and memory interface chips, the probe connections provide a display of the activity on the address lines located on the memory interface chip. The logic state analyzer controls

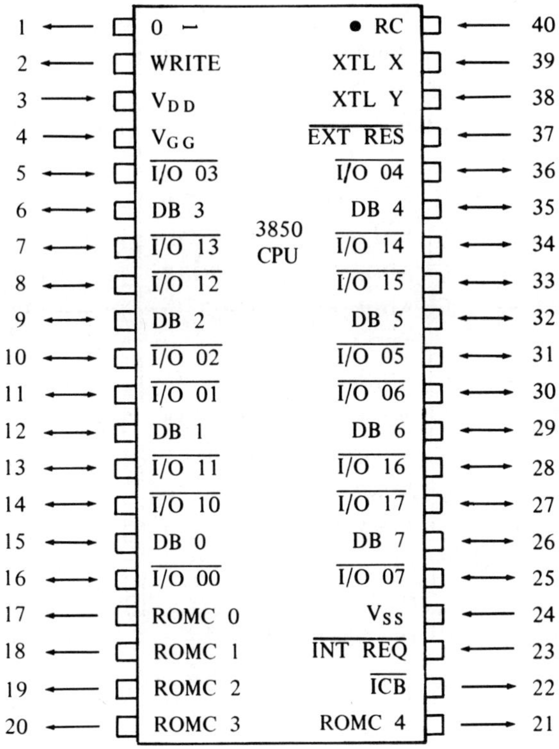

Appendix III · 251

```
  1 ──→ ┤ Φ ─             • RC  ├──→ 40
  2 ──→ ┤ Φ WRITE        XTL X  ├──→ 39
  3 ──→ ┤ V_DD           XTL Y  ├──→ 38
  4 ──→ ┤ V_GG          EXT RES ├──→ 37
  5 ──→ ┤ I/O 03         I/O 04 ├──→ 36
  6 ──→ ┤ DB 3           DB 4   ├──→ 35
  7 ──→ ┤ I/O 13  3850   I/O 14 ├──→ 34
  8 ──→ ┤ I/O 12   CPU   I/O 15 ├──→ 33
  9 ──→ ┤ DB 2           DB 5   ├──→ 32
 10 ──→ ┤ I/O 02         I/O 05 ├──→ 31
 11 ──→ ┤ I/O 01         I/O 06 ├──→ 30
 12 ──→ ┤ DB 1           DB 6   ├──→ 29
 13 ──→ ┤ I/O 11         I/O 16 ├──→ 28
 14 ──→ ┤ I/O 10         I/O 17 ├──→ 27
 15 ──→ ┤ DB 0           DB 7   ├──→ 26
 16 ──→ ┤ I/O 00         I/O 07 ├──→ 25
 17 ──→ ┤ ROMC 0         V_SS   ├──→ 24
 18 ──→ ┤ ROMC 1        INT REQ ├──→ 23
 19 ──→ ┤ ROMC 2         ICB    ├──→ 22
 20 ──→ ┤ ROMC 3        ROMC 4  ├──→ 21
```

PIN NAMES	DESCRIPTION	TYPE
DB 0 - DB7	Data Bus	Bidirectional
I/O 00 - I/O 07	I/O Port Zero	Input/Output
I/O 10 - I/O 17	I/O Port One	Input/Output
ROMC 0 - ROMC 4	Control Lines	Output
RC	RC Timing Input	Input
XTL - X	Crystal Inputs	Input
XTL - Y	External Clock Inputs	Input
EXT RES	External Reset	Input
Φ WRITE	Clocks	Output
ICB	Interrupt Control Bit	Output
INT REQ	Interrupt Request	Input
V_DD V_SS V_GG	Power	Input

FIGURE III-1 Central processing unit chip pin assignments.

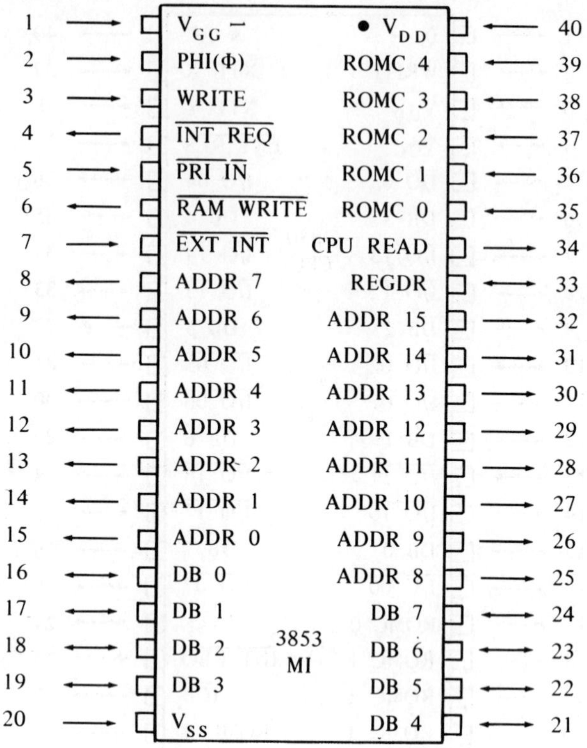

```
 1 →  ☐ V_GG         • V_DD  ☐ ← 40
 2 →  ☐ PHI(Φ)       ROMC 4  ☐ ← 39
 3 →  ☐ WRITE        ROMC 3  ☐ ← 38
 4 ←  ☐ INT REQ      ROMC 2  ☐ ← 37
 5 →  ☐ PRI IN       ROMC 1  ☐ ← 36
 6 ←  ☐ RAM WRITE    ROMC 0  ☐ ← 35
 7 →  ☐ EXT INT      CPU READ ☐ → 34
 8 ←  ☐ ADDR 7       REGDR   ☐ ↔ 33
 9 ←  ☐ ADDR 6       ADDR 15 ☐ → 32
10 ←  ☐ ADDR 5       ADDR 14 ☐ → 31
11 ←  ☐ ADDR 4       ADDR 13 ☐ → 30
12 ←  ☐ ADDR 3       ADDR 12 ☐ → 29
13 ←  ☐ ADDR 2       ADDR 11 ☐ → 28
14 ←  ☐ ADDR 1       ADDR 10 ☐ → 27
15 ←  ☐ ADDR 0       ADDR 9  ☐ → 26
16 ↔  ☐ DB 0         ADDR 8  ☐ → 25
17 ↔  ☐ DB 1         DB 7    ☐ ↔ 24
18 ↔  ☐ DB 2   3853  DB 6    ☐ ↔ 23
                MI
19 ↔  ☐ DB 3         DB 5    ☐ ↔ 22
20 →  ☐ V_SS         DB 4    ☐ ↔ 21
```

PIN NAME	DESCRIPTION	TYPE
DB 0 - DB 7	Data Bus	Bidirectional
ADDR 0 - ADDR 15	Address	Output
Φ WRITE	Clocks	Input
INT REQ	Interrupt Request	Output
PRI IN	Priority in Line	Input
RAM WRITE	Write Line	Output
EXT INT	External Interrupt Line	Input
REGDR	Register Drive Line	Input/Output
CPU READ	CPU Read Line	Output
ROMC 0 - ROMC 4	Control Lines	Input
V_{SS} V_{DD} V_{GG}	Power Lines	Input

FIGURE III-2 Memory interface chip pin assignments.

are set in the same general manner as noted in Appendix I. If the program is not looping or cycling through the selected address, select "Single," press "Reset," and start the system. The first time that the system passes through the selected trigger state, the display will be generated and stored.

A portion of a start-up program is illustrated in Figure III-4. Proper operation is confirmed by a comparison between real-time state analysis and the start-up program. With the Q0 probe connected to CPU READ, Q0 set HI, and triggering on address 0000, an address is incremented with every memory fetch. Notice that the address sequence branches after address 000D, where it loops back to clear all F8 scratch-pad registers, and then the program is reentered at address 000E. The "Paging" technique is used to determine whether this loop is completed. Reset the logic state analyzer trigger-word switches to address 000D, and restart the system. The display then begins at address 000D, and the next 15 addresses are displayed. Some addresses in the start-up program involve multiple-word instructions. These may be observed on the display as redundant consecutive addresses by placing the Q0 switch in the OFF position.

If a tabular display is not presented in the previous step, it means that the system did not access the selected address, and the No-Trigger light will be on. To find where the system is residing in the program, switch to "map" (see Figure III-5). Use the Trigger-Word switches to move the cursor to encircle one of the dots on the screen. Switch to "expand," and make the final positioning of the cursor. The No-Trigger light will now go out, and switching back to Table A displays the 16-bit address around that point. When program deviations are found, the reason may be as simple as a program error or as complicated as a hardware failure on the data bus, control bus, or other command lines. Additional input channels are needed at this point by the troubleshooter. By combining the 1600A and 1607A, the display and trigger capability can be expanded to 32 bits wide, allowing the 16-bit address, 8-bit data bus, and eight other active command signals to be viewed simultaneously.

As an amplification of the previous example, it is possible to investigate all the activity on the address and data buses plus the CPU READ control line. Operating the logic state analyzer system as noted previously, the display shows all activity on the address bus as well as on the data bus during the start-up program. In addition, the CPU READ control line is monitored to permit confirmation of system performance. Figure III-6 shows multiple-word instruction addresses and at the same time shows acitivity on the address bus during the execution

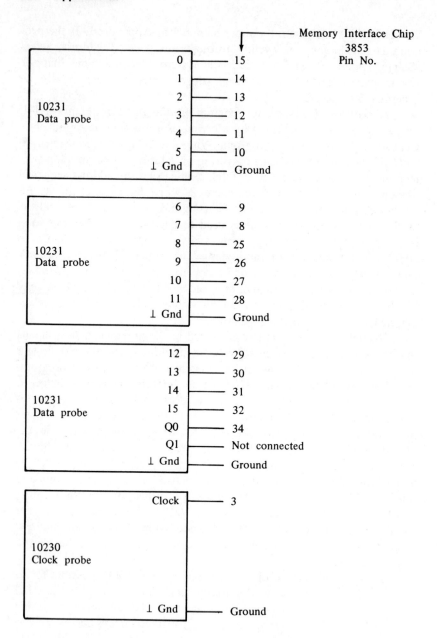

FIGURE III-3 Probe connections.

Appendix III · 255

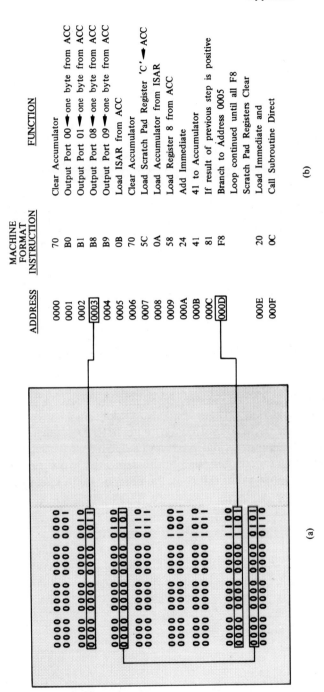

ADDRESS	MACHINE FORMAT INSTRUCTION	FUNCTION
0000	70	Clear Accumulator
0001	B0	Output Port 00 ← one byte from ACC
0002	B1	Output Port 01 ← one byte from ACC
0003	B8	Output Port 08 ← one byte from ACC
0004	B9	Output Port 09 ← one byte from ACC
0005	0B	Load ISAR from ACC
0006	70	Clear Accumulator
0007	5C	Load Scratch Pad Register 'C' ← ACC
0008	0A	Load Accumulator from ISAR
0009	58	Load Register 8 from ACC
000A	24	Add Immediate
000B	41	41 to Accumulator
000C	81	If result of previous step is positive
000D	F8	Branch to Address 0005
		Loop continued until all F8 Scratch Pad Registers Clear
000E	20	Load Immediate and
000F	0C	Call Subroutine Direct

(b)

FIGURE III-4 System response to a start-up program.

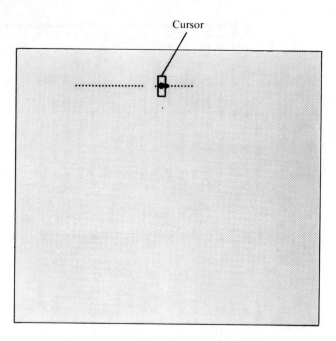

FIGURE III-5   Map display shows entire system activity

of the instruction. At address 0004, the data bus indicates that I/O Port 08 is addressed. Then, contents of the accumulator are sent to the I/O Port. In the next instruction, CPU READ is high, the data bus shows that the instruction has been completed at instruction code B9, and the program counter has been incremented to the next address. In a similar manner, each address in the program may be analyzed.

Appendix III · 257

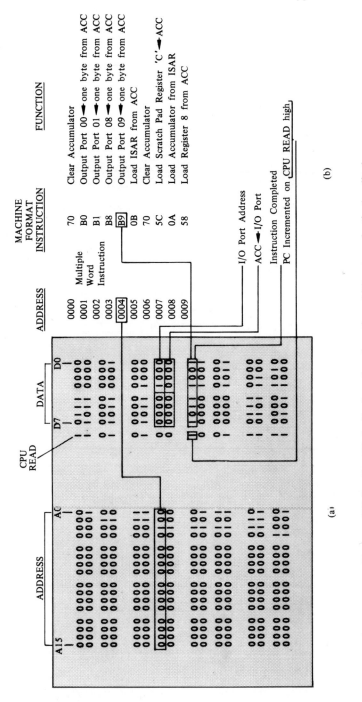

Figure III-6 Qualified display showing the ability to selectively display only desired address data.

GLOSSARY

ABBREVIATIONS FOR LOGIC FORMS:

    CML: Current-mode logic
    CTL: Complementary transistor logic
    DCTL: Direct-coupled transistor logic
    DTL: Diode-transistor logic
    ECL: Emitter-coupled logic
    RCTL: Resistor-capacitor-transistor logic
    RTL: Resistor-transistor logic
    TTL: Transistor-transistor logic

ACCESS TIME. Time required in a computer to move information from the memory to the computing subsystem.

ACCUMULATOR. The register (row of flip-flops) that displays an answer such as in an adder or subtracter configuration; a register in an arithmetic unit for performance of arithmetical and logic functions such as addition and shifting.

ACTIVE ELEMENTS. Those components in a circuit that provide gain, or control DC flow; diodes, transistors, SCR's, and so on.

ADD CONTROL. The "halt" command that is used to control (or stop) the addition process in an adder subsystem.

ADDEND. The number that is to be added to another number in the process of addition.

ADDEND REGISTER. The register (row of flip-flops) in which the number to be added is entered in an adder subsystem.

ADDER. Switching circuits that combine binary bits to generate the sum and carry of these bits. An adder takes the bits from the two binary numbers to be added (addend and augend) plus the carry from the preceding less-significant bit and generates the sum and carry.

ADDER, FULL. The complete logic circuitry that generates both a sum and a carry output.

ADDER, HALF. The logic circuitry that generates only a sum output and not a carry output; also called a *sum gate*.

ADDER, LOGIC. A wired configuration of logic gates and/or flip-flops that will perform binary addition.

ADDER, NON-GATES BINARY. A wired configuration of flip-flops only, which will perform addition; i.e., which will allow a binary number to be added in one register and the answer to be displayed in a second register.

ADDER, SHIFT. A wired configuration of flip-flops and logic gates that consist of one full adder and three shift registers. Two shift registers shift the two numbers to be added through the full adder, and the answer is shifted out in the third-shift register.

ADDITION IDENTITY. A logic expression that defines a sum, a carry, or both, for two or more binary numbers.

ADDRESS. A location, either a name or number, where information is stored in a computer. To select or pick out the location of a stored information, set for access.

ALPHANUMERIC. A combination of alphabetic characters and digital numbers.

ANALOG. Representing something else. Thus, a meter movement that indicates a voltage value on a scale. A continuous variation from one value to another; thus, "nondigital".

AND GATE. An electronic circuit that forms a logic gate whose output is a 1 only when all of its inputs are 1. The gate output is 0 for all other combinations of inputs.

AND LOGIC OPERATION. The logic operation denoted by a dot (.), which answers the question "Are all the facts TRUE?" or, "Are all the inputs 1?"

ANDING. The process of applying signals to an AND function.

ANODE. The lead on a diode or other device that receives positive voltage, as opposed to the other lead, the cathode, which receives negative voltage.

ARCHITECTURE. The construction or making of a system.

Glossary · 261

Astable Multivibrator.   A free-running pulse generator.

Asynchronous Inputs.   Those terminals of a flip-flop that can affect the output state of the flip-flop independently of the clock. Also called *set, preset, reset, DC set,* and *clear,* or *clear inputs.*

Asynchronous Operation.   Usually, an operation that is started by a completion signal from a previous operation. It then proceeds at maximum speed until finished and then generates its own completion signal.

Augend.   The number to which another number will be added.

Bar.   Denotes the inverse, or complement, of a function. Example: The inversion of A is written $\overline{A}$, and is read A Bar, or A Not.

BCD.   An abbreviation for Binary Coded Decimal.

Binary.   Two-valued. Thus, the value may be True and False, 1 and 0, On and Off, and so on; one and only one of the two values is always present. Also, a number system to the base 2, which represents any number by combinations of the binary digits 1 and 0.

Binary Addition.   A process of obtaining the binary sum of two binary numbers.

Binary Coded Decimal.   (See BCD)

Binary Complement.   That binary number that has the 1's and 0's of its digits interchanged with those of the original binary number. The sum of a binary number and its complement is a third binary number whose digits are all 1's.

Binary Division.   A process of dividing a binary number by another binary number to obtain a binary quotient.

Binary Fractional.   All binary digits to the right of the binary point.

Binary Input.   A binary 1 or 0 that is fed to a logic gate input of flip-flop input in order to perform a logic function or flip-flop triggering operation; any pin on a flip-flop or logic gate that can receive a binary signal.

Binary Logic.   Digital logic elements that operate with two distinct states. The two states are represented in calculators and computers by two different voltage levels.

Binary Multiplication.   A process of multiplying two binary numbers together to obtain a binary product.

Binary Number.   A digit that represents a power of 2 and can only be a 1 or 0; a number of two or more digits that represents powers of 2, and each digit is a 1 or a 0.

Binary Output.   Any terminal of flip-flop or logic gate that generates a binary 1 or 0 logic-level signal resulting from one or more input signals. Any terminal on a pulse generator that generates a pulse-voltage signal.

BINARY POINT. Similar to a decimal point (.), except that it is used to denote a binary fractional.

BINARY SUBTRACTION. A process of subtracting a binary number from another binary number to obtain the binary difference.

BINARY SYSTEM. The number system to the base 2.

BISTABLE MULTIVIBRATOR. A flip-flop.

BIT. A bit is one character of a digital word. It can be either a 1 or a 0. The position of the bit in the word usually determines its significance. The first bit may be a sign bit, with 0 representing a $+$ sign and 1 representing a $-$ sign.

BLANKING INPUT. A control input for decoding circuits.

BLOCK. A specified number of digital characters. The block may be of any size, but block size is usually uniform within a system or subsystem.

BOOLEAN APPROACH. To impose the condition that all logic statements, reasons, conclusions, facts, and so on, are either True or False.

BOOLEAN EQUATION. Mathematic expression of logic relationships in terms of True or False.

BORROW. The binary 1 that is removed from the adjacent column to the left by the subtraction of binary 1 from binary 0, leaving 1 in the column that was subtracted.

BRACES. Symbols { } that denote logic grouping, along with parentheses and brackets.

BRACKETS. Symbols [ ] that denote logic grouping, along with parentheses and braces.

BUFFER. An amplifier that processes digital information. A temporary storage location for a four-bit binary number.

BYTE. A group of bits (usually 8 bits).

CANCEL. To remove a binary 1 from a flip-flop; to turn a flip-flop readout light off; the process of removing a binary number from a register.

CARRY. The binary 1 that is generated in the next column to the left by the addition of a pair of binary 1's, leaving 0 in the column that was added.

CARRY FEEDBACK. A carry, or binary 1, that is generated in a column other than the next column to the left; the carry of the last binary number at the extreme left of a register that is fed back into the first binary number (extreme right) of the same register.

CATHODE. The lead on a diode or other device that receives negative voltage, as opposed to the other lead, the anode, which receives positive voltage.

CHARACTER. A combination of bits having a specified assigned meaning;

one element of a code. Characters are usually composed of 4, 5, 6, 7, or 8 bits each.

CLEAR. A circuit is said to be cleared when it is set to all 0's.

CLEAR INPUT. A clear input generates a clear function.

CLOCK. A pulse train of known frequency and waveform, used to synchronize the operation of circuits within a specific device or system. A device may have several different clocks, but each clock bears a defined relationship to all other clocks.

CODE. A means of representing information in digital form by assigning a specific pattern of bits to each item of information. Each information-item/bit-pattern pair is known as a character. An example is binary coded decimal assignment of four-bit binary representations to the decimal digits.

CODER. A device for converting data from one notation system to another.

COMMAND GENERATOR. A binary counter used in conjunction with AND gates to generate a repetitive series of command pulses. An n-bit counter can generate up to 2^n repetitive command pulses.

COMPLEMENT, BINARY. That binary number which has the 1's and 0's of its digits interchanged from those of the original binary number.

CORE. A particle of magnetic material used to store digital data, usually but not always in the shape of a toroid. By extension, an array of cores comprising a memory.

COUNTER. A device capable of changing states in a specified sequence upon receiving appropriate input signals. The output of the counter indicates the number of pulses that have been applied.

COUNTER, BINARY. An interconnection of flip-flops having a single input so arranged to enable binary counting.

COUNTER, BINARY CODED DECIMAL (BCD). A binary counter with logic gating that will count up to binary "nine" (1001), and then reset itself to 0000.

COUNTER, DOWN. A counter that starts from any specified number and decreases its value by 1 with each pulse signal input; it "counts backward."

COUNTER, GATES "UP–DOWN." A single-register binary counter that can be controlled by logic gating so as to count up or count down at the proper command.

COUNTER, RING. A special form of counter, with simple wiring. It forms a loop, or circuits of interconnected flip-flops so arranged that only one is 0 and that as input signals are received, the position of the 0 state moves in sequence from one flip-flop to another around the loop until they are all 0, after which the first one goes to 1, and this moves in sequence from one flip-flop to another until all are 1. It has $2 \times n$ possible counts where n is the number of flip-flops.

COUNTER, UP–DOWN. A binary counter wired with extra logic gates so that the free-running counter will count alternately up and then switch automatically to count down after resetting from the up count.

DATA. Denotes facts, numbers, letters, symbols, in binary bits presented as voltage levels in a computer. Binary data can consist only of 0 and 1.

DATA PROCESSOR. Any digital device; usually associated with devices not used for mathematical computation.

DECIMAL SYSTEM. Our conventional number system, which employs the digits 1, 2, 3, 4, 5, 6, 7, 8, 9, and 0.

DECODER, BINARY TO DECIMAL. A configuration of logic gating that will convert any binary number to a decimal number; usually employed with a binary coded decimal counter.

DECOUPLE. To block out any electronic noise or interference that is generated by nearby circuits, by the power source (power supply), or by external noise sources.

DELAY. The slowing up of the propagation of a pulse either intentionally to prevent inputs from changing while clock pulses are present, or unintentionally as caused by transistor rise and fall time pulse-response characteristics.

DEMORGAN'S THEOREM. This theorem states that the inversion of a series of AND implications is equal to the same series of inverted OR implications, or, the inversion of a series of OR implications is equal to the same series of inverted AND implications.

$$\overline{A.B.C} = \overline{A} + \overline{B} + \overline{C} \qquad \overline{A+B+C} = \overline{A}.\overline{B}.\overline{C}$$

DIAGRAMS, LOGIC. Drawings that illustrate how flip-flops and logic gates must be connected to perform specific computer functions; the use of symbols to represent flip-flop, logic-gate, and pulse-generator electronic circuits without drawing out the full electrical schematic each time.

DIFFERENCE. The answer that results from the subtraction of a number from another number; the binary digit that represents the subtraction of one binary digit from another binary digit, exclusive of the "borrow."

DIGIT. A single number.

DIGIT, BINARY. A single binary number; a 1 or a 0.

DIGIT, LEAST SIGNIFICANT. The digit at the extreme right of any number.

DIGIT, MOST SIGNIFICANT. The digit at the extreme left of any number.

DIGIT, SENSE. The use of an OR gate to determine whether or not a number is present in a register.

DIGIT SENSE LOGIC. A gated logic circuit (usually a multi-input OR gate) that determines whether any numbers (or digits) are present in a register.

DIGITAL. Representation of a number in discrete terms of On or Off. The flip-flop register is a digital representation of a binary number.

DIGITAL CIRCUIT. (Binary Circuit); a semiconductor configuration that operates as a switch.

DIP. Abbreviation of dual-in-line package.

DIRECT OUTPUT DISPLAY. The use of a flip-flop to read out the output from a logic gate by wiring that output of the logic gate directly into the flip-flop.

DISCRETE CIRCUITS. Electronic circuits comprising separate, individually manufactured, and assembled diodes, resistors, transistors, capacitors, and other specific electronic components.

DIVIDEND. A number to be divided (in the arithmetical process of division).

DIVIDER, LOGIC. A configuration of gated logic circuits and registers that will divide one number by another.

DIVISOR. A number by which a second number is divided (in the arithmetical process of division).

DOUBLING, BINARY. The process of adding an extra 0 to the left of a given binary number. This added 0 results in doubling the number.

DOWN COUNTER. A counter that starts from a specified number and decreases its value by 1 with each pulse signal input; thus, the circuit "counts backward."

DOWN-SWING, VOLTAGE. A change in output voltage from a positive value to zero; or, a change from a negative value to zero (in positive logic).

DRIVER. An element coupled to the output stage of a circuit to increase its power capability or fanout.

DTL. Diode-transistor logic; employs diodes as switches, with transistors as inverting amplifiers.

DUTY CYCLE. The ratio of on-time to total cycle time of a pulse waveform.

ENABLE. To permit an action by application of appropriate signals (generally a logic 1, in positive logic) to the appropriate input.

END-AROUND. Transfer of a pulse command from the last flip-flop in a register to the first flip-flop.

ENTER. To place a binary 1 in a flip-flop or flip-flops.

EQUALITY, LOGIC. Defines two or more logic expressions, equations, or facts to be identical.

FACT, LOGIC. A statement (which may be true or false) that is used in a reasoning process to arrive at a true or false conclusion.

FALL TIME. The decay time of the trailing edge in a pulse waveform.

FALSE. Statement for a 0 in Boolean algebra.

FAN-IN. Total number of inputs to a particular gate or function.

FAN-OUT. Total number of loads connected to a particular gate or function.

FILE. One or more records of information arranged in a sequence and stored for future use.

FLIP-FLOP. A circuit with two and only two stable states.

D flip-flop: Delay binary; the output shows the input signal at the next clock pulse (one clock pulse delayed).

J-K flip-flop: Binary with synchronous set and reset inputs.

RS flip-flop: (Latch) Binary with set and reset inputs, and the restriction that both inputs shall not be energized simultaneously.

RST flip-flop: Binary with the features of an RS and a T flip-flop.

RS/T flip-flop: An RS flip-flop that may be connected to operate in the toggling mode.

T flip-flop: (Toggle) Binary with a synchronous T input; if the T input is high, the flip-flop will toggle (change state) synchronously.

FULL ADDER. The complete logic circuitry that generates both a sum and a carry output.

FULL SUBTRACTER. The complete logic circuitry that generates both a difference and a borrow output.

FUNCTION, LOGIC. An expression that contains one or more logic operators, indicating logic operation(s) to be performed; a specific logic operation such as AND or OR.

GATE. The simplest logic circuit; its output voltage will be high or low depending on the states of the inputs, and the type of gate that is employed.

HALF-ADDER. A switching circuit that combines binary bits to generate the sum and carry; it can only take in the two binary bits to be added and generate the sum and carry.

HANG-UP. The inability of a flip-flop to be triggered from a pulse command.

HIGH. (See binary logic)

HOLDING TIME. The period of time that the input states must remain after activation of the clock input.

INHIBIT. To prevent an action (opposite of enable).

INTEGRATED CIRCUIT (IC). The physical realization of a number of electrical elements inseparably associated on or within a continuous body of semiconductor material to perform the functions of a circuit.

INVERTER. A device or element to complement a Boolean function.

J-K FLIP-FLOP. (See Flip-flop)

KEYBOARD. A device for encoding information by depression of various key switches.

KILOBIT. One thousand binary bits.

LABELS. Identifying statements or numbers that are used to describe flip-flop or logic-gate relative positions.

LATCH. Usually, a feedback loop in a symmetrical digital circuit, such as a flip-flop, for retaining a state; synonym for a set-reset flip-flop.

LAW, LOGIC. A relation that is proved or assumed to hold between other logic expressions (expressed as a logic equality).

LEAKY. A poor front-to-back ratio in a semiconductor junction.

LEAST SIGNIFICANT BIT. The bit in a number that is least important (has the least weight).

LINEAR CIRCUIT. A circuit whose output is an amplified version of its input, or whose output is a continuous predetermined variation of its input.

LOGIC. The science of reasoning; making use of known facts to reason out a conclusion; the arrangement of electrical circuits that applies the defined rules or functions to input signals to produce output information.

LOGIC DIAGRAMS. Drawings that show how flip-flops and gates must be connected to perform specific computer functions.

LOGIC GATE. An electronic circuit that performs a logic operation, such as AND or OR.

LOGIC LEVELS. One of two possible states; 0 or 1.

LOGIC SWING. The voltage difference between logic-high and logic-low levels.

LOW. (See binary logic)

LSB. Least significant bit; the lowest weighted digit in a binary number.

LSI. Abbreviation for large-scale integration; a chip containing more than 100 gates.

MASTER-SLAVE. A binary element containing two independent storage states, with a definite separation of the clock function to enter information into the master and to transfer it to the slave.

MATRIX. A group of elements such as circuit components or information items arranged in a defined relationship in such a manner that the intersection of two or more inputs produces a unique output.

MEMORY. Any device used to store information; frequently a matrix of magnetic cores, or magnetic tapes, drums, or disks.

MSI. Abbreviation for medium-scale integration; a chip having from 10 to 100 gates.

MULTIPLEX. Commutate; to sequentially connect a central unit or system to one of several channels.

MULTIPLEXING.. The process of combining the data from a number of sources into one flow of data. The reverse process of sorting out multiplexed data is called *demultiplexing*.

NAND. Logic function that produces the inverted AND function.

NEGATION. Employs an overhead bar to denote not-A when placed over logic fact A.

NEGATION, DOUBLE. A process of inverting twice or inverting the negative of a logic fact to revert it to the original fact.

NEGATIVE-EDGE GATING. Circuit response as the control signal goes from high to low.

NEGATIVE LOGIC. Logic in which the more negative voltage represents the 1 state; the less negative voltage represents the 0 state.

NOR GATE. An electronic circuit that forms a logic gate and whose output is a 1 only when all of its inputs are 0.

NOT. A Boolean logic operation denoting negation; an inverter.

OPEN-COLLECTOR OUTPUT. A TTL gate with only one output transistor, instead of a two-transistor totem-pole configuration.

OR. Logic operation that produces a 1 at the output if at least one input is a 1.

OUTPUT, BINARY. Any pin of a flip-flop or logic gate that generates a binary 1 or 0 voltage resulting from one or more input signals.

OVERFLOW. The carry or borrow output generated by the last flip-flop at the end of a register.

PARALLEL. A technique for processing a binary data word that has more than one bit; all bits are acted upon simultaneously.

PARALLEL ADDER. A technique for addition in which the two multibit numbers are presented and added together simultaneously.

PASSIVE ELEMENTS. Resistors, inductors, or capacitors; elements without gain.

POSITIVE-EDGE GATING. A circuit that responds as the control signal goes from low to high.

POSITIVE LOGIC. Logic operations in which the more positive voltage represents the 1 state.

PRESET. An input like the Set input, which operates in parallel with the Set.

PULL-DOWN RESISTOR. Usually, a resistor connected to ground or to a negative voltage, as from the base of a transistor to a negative voltage point.

PULL-UP (ACTIVE). A transistor that replaces the pull-up resistor to obtain low output impedance and low power consumption.

PULL-UP RESISTOR. Usually, a resistor connected to the positive supply, as from V_{cc} to the output collector terminal.

Q OUTPUT. The reference output of a flip-flop. If the Q output is in the 1 state, the flip-flop is said to be in its 1 state.

$\overline{Q}$ OUTPUT. The second output of a flip-flop; its logic level is always opposite to that of the Q output.

RACE. A condition that exists whenever a signal propagates through two or more memory elements during one clock period.

REGISTER. A storage device for binary data generally used to store numbers for arithmetical operations or their result.

RESET. Setting a flip-flop to the Q=0 state; also applies to any circuit when driven to its normal starting condition.

RISE TIME. The time required for the leading edge of a waveform to proceed from its 10% to its 90% of maximum amplitude points.

SERIAL DATA. The data are available as a series of bits occurring one after the other in a single file.

SET. Placing a flip-flop in its Q=1 state.

SHIFT. The process of moving data from one place to another.

SHIFT REGISTER. A wired configuration of flip-flops that will shift all 1's in a binary number either one position to the left or one position to the right with each pulse command.

SIGNATURE. The particular reference signal of a given circuit.

SINK LOAD. A load with current flow out of its input; a current load must be driven by a current sink.

SOURCE LOAD. A load with current flow out of its input; a current load must be driven by a current sink.

STATE. Refers to the condition of an input or output of a circuit concerning whether it is a logic 1 or a logic 0.

STROBE. An input to a counter or register that permits the entry of parallel data asynchronously.

SYNCHRONOUS. Operation of a switching network by a clock-pulse generator. All circuits in the network switch simultaneously. All actions take place synchronously with the clock.

TOGGLE. To change a binary storage element to its opposite state.

TRIGGER. The input pin on a flip-flop; a pulse command signal will cause the flip-flop to change state; a timing pulse used to initiate the transmission of logic signals through the appropriate signal paths.

TROUBLESHOOT. To search for errors or malfunctions, in order to correct them.

TRUE. A true condition is the statement for a logic 1 in Boolean algebra.

TRUTH TABLE. A tabular list with all possible input logic combinations and the resulting logic output for all these combinations.

UNLOAD. To remove information in massive quantities.

UP COUNTER. A binary counter that starts from zero and increases its value by 1 with each pulse signal input.

UP-SWING, VOLTAGE. A waveform change in level from a lower value to a higher value.

VLSI. Abbreviation for very-large-scale integration; a chip containing more than 1,000 gates.

SPECIAL ACKNOWLEDGMENTS

Design of Digital Computers. Gschwind, McClusky, Springer Verlag, 1975, pp. 122–40
"Computer Generated Fault Diagnostics." W. J. Haydamack and McClure. p. 180 Proc. NEPCON-WEST, Feb. 11–13, 1975, Anaheim, California.
"Generation of Fault Isolation Tests for Logic Networks." W. J. Haydamack, IEEE Intercon 75, Conference Record, Session 21, pp. 21–23. April 8–10, 1975.

Contributors:

Bill Haydamack, Ken Parker, Hal Brook, Homer Tsuda, Bob Aiken, Juris Blukis, and Mark Baker.

INDEX

Access register, 24
Accumulator, 24, 33, 42
Activity, pulse, 83
Adapter, interface, 215
Adder, 31, 32
Addition, binary, 29
Addresses, 26
Adder
    full, 31
    half, 30
    parallel, 31
    serial, 32
Addition, binary, 29
Aggravation, 119
ALD, 134
Algorithm, 31, 40, 220
ALU, 161
Analog component, 70
Analysis, electrical, 104
Analyzer
    display, 110
    logic, 140
AND gate, 1
A NOT, 4

Architecture, computer, 21
Arithmetic unit, 21, 24
Asynchronous clock, 80
Attack, sequential, 114
Augend, 33

Backplane, 157
Bad level, 56, 75
Bar, reset, 111
BCD, 44
Bed-of-nails fixture, 152
Bias, quiescent, 220
Bilevel system, 107
Binary
    addition, 29
    coded decimal, 44
    division, 44
    multiplication, 39
    subtraction, 33
    voltmeter, 61
    word, 26
Bit error rate, 101

Board
   circuit, 65
   modeling, 185
   multilayer, 76
   reference, 65
Bond, open, 69, 86
Boolean logic, 2
Buffer, reader, 120
Burst, noise, 139
Bus, interface, 107

Calculator, 19
Capacitive pull-up, 210
Carry
   end-around, 34, 36
   ripple, 32
Chip, memory, 6
Clear, 19
   line, 211, 219
Clip, logic, 61
Clock, 19, 57
   asynchronous, 80
   signal, 83
Clockout, 206
Clutter, 80
CMOS, 76
Code
   instruction, 26
   operation, 26, 131
Coded instructions, 26
Coincidence, 51
Command, 80
Comparator, logic, 61
Complement, 17, 33
   operations, 35
Computational sequence, 164
Computer, digital, 21
Connectors, edge, 153
Counter, decade, 83
Critical race, 179, 184
Cross-check, 90

DAC, 159

Data, 4
   clockout, 206
   management, 24
   processing, 24
   word, 26
Dead node, 66
Debugging, 104
Decade counter, 83
Decision making, 4
Decoder circuits, 182
D-latch, 162
Delay, 213
   macro, 162
   micro, 162
   negative, 148
Delayed transition, 193
Dictionary, fault, 215
Digital
   architecture, 21
   computer, 21
   flow, 106
   plotter, 96
   test system, 216
   words, 26
Discrete logic, 65
Disk memory, 101
Double transition, 225
Driver, parallel, 223
DTL, 56
Dual function, 11

ECL, 65
Edge connector, 153
Electrical analysis, 104
Element
   logic, 171
   memory, 205
   mythical, 171
   primitive, 196
EMI, 106
Enabled state, 170
Enabling input, 83
End-on-trigger, 103
Equation, logic, 2

Equivalent fault, 197
Event, 14
   checking, 51
EXCLUSIVE OR, 13
   symbol, 14
Exercise loop, 117

Fail, stop on, 200
Failure path, 219
Fanout points, 201
Fault
   dictionary, 215
   equivalent, 197
   grouping, 194
   internal, 203
   location, 217
   machine, 197
   marginal, 69
   signatures, 198
   sites, 201
File
   fault signature, 214
   node state, 214
   pattern, 214
   topology, 214
Filter, pulse-width, 149
Flag, 96
   signals, 98
Flip-flop, 16
   J-K, 19
Flow diagram, 121
Front-panel milking, 82
Full
   adder, 31
   subtracter, 34
Functional testing, 102

Gate, 1
   AND, 1
   EXCLUSIVE OR, 13
      symbol, 14
   NAND, 10

   NOR, 2
   XOR, 29
Generator, pattern, 207
Glitch, 78, 106
Group fault, 194
Guided probe, 226

Half
   adder, 31
   subtracter, 34
Handshake signals, 107
Hardware validation, 227
Hazard, 179, 186
High, logic, 47
HiNIL, 65
Hold
   mode, 91
   time, 104
HTL, 65
Hybrid device, 196
Hysteresis, 60

IC failures, 103
ILD, 134
Index reference, 109
Indexing, 107
Indicator panel, 115
Inhibit signal, 91
Initialization, 21, 204, 213
Injection, signal, 86
I/O, 120
Integrated circuit, 4, 56
Interchange, symbol, 12
Interface
   adapter, 215
   bus, 107
   indicator, 107
Intermittent, 89
Internal faults, 200, 203
Inverter, 3, 11

J-K flip-flop, 19
Jitter, 142
Jump instruction, 104, 118

Key signals, 82

Language processor, 215
Latch, 16
    D, 162
    RS, 14
Latency, 179
LED, 64
Logic, 1
    analyzer, 7, 80
    circuit, 8
    clip, 61
    comparator, 63
    element, mythical, 171
    equation, 2
    high, 47
    level, 17
    low, 47
    packages, 57
    probe, 47, 55
        multifamily, 76
        pulser, 54
    redundant, 208
    sequential, 218
    state response, 182
LSB, 32

Machine language, 26
Macro delay, 162
Mapping, 86
Marginal faults, 69
Memory, 2, 14
    chip, 6
    primitives, 162
Microprogramming, 24
Micro delay, 162

Microprocessor, 80
Minuend, 34
Misprobe, 220
MLD, 134
Model
    unbuffered, 224
    verification, 182
Module, 4
Monitoring test, 51
MOS, 65
MSB, 32
MSI, 197
Multifamily probe, 76
Multilayer boards, 76
Multiple
    fault sites, 201
    inputs, 10
Mythical element, 171

NAND, 10
    gate, 18
Negative
    delay, 148
    time, 142
NMOS, 76
Node
    dead, 66
    state file, 214, 226
Noise
    burst, 139
    system, 106
Noncritical race, 184
NOR, 10
    gate, 18
NOT A, 4

Octal-coded ROM, 103
Open bond, 69, 86
Operand, 24, 26
Operator-directed probe, 226
OR gate, 2
Oscillation, 182

Parallel
    adder, 31
    driver, 223
Parameter response, 206
Path sensitization, 203
Pattern
    file, 214
    generator, 207
    selection, 203
    sensitization, 209
    set, 203, 218
PCB, 164
Peripheral equipment, 24
Pin diagram, 4
Possible detect, 204
Post processor, 215
Preset, 19
PRBS, 107
Primitive elements, 196
Probe
    algorithm, 220
    guided software, 226
    multifamily, 76
    operator-directed, 226
    sequence, 223
Program, 26
    source-language, 226
Pull-up
    capacitor, 210
    resistor, 189
Pulse
    activity, 83
    repetition rate, 48
    rise and fall, 48
    single, 49
    stretching, 163
    trains, 52
    transfer, 57
    waveshapes, 52
    width, 56

Quiescent bias, 220
Quad
    NAND gate, 165
    version, 183

Qualifier units, 96

Race, 18
RAM, 95
Real time, 175
Reconvergent path, 208
Rectangular waveform, 53
Redundant logic, 208
Register, 88
Reinitialization, 182
Relay logic, 65
Reset, 17, 57
Restart, 80
Ripple carry, 32
Rise time, 189
ROM, 22, 95
Routine, 26
RS latch, 14
RTL, 76

Screen clutter, 80
Sensitizing, 170
Serial adder, 32
Set input, 16
Shift, 57
    register, 32, 77
Signal injection, 86
Simulator, 211
    two-valued, 177
    three-valued, 177
Single stepping, 86
Software, 175
Start, 57
States, internal, 215
Static high, 71
Steering circuit, 70
Stop on fail, 200
Storage mode, 49
Strobe, 91
Subtrahend, 34
Sum register, 32
Symbol interchange, 12

Test system, 216
Threshold, 69
Time
    hold, 104
    negative, 142
    setup, 104
Timing
    diagram, 5, 8
    skew, 78
Tolerance, 53
Totem pole, 69
Tracking test, 90
Trade-offs, 202
Transcribe, 26
Transients, 78
Transition time, 104
Triggered sweep, 177
Tristate
    configuration, 167
    inverter, 169
Truth table, 1, 7
TTL, 56

Unauthorized pulse, 111
Unbuffered model, 224
Unique
    detection, 201
    trigger, 142
Unit delays, 164
Unmodeled signature, 224
Unprobable nodes, 171
UUT, 156

Validation, 227
Verification, model, 182

Wait loop, 102
Waveform
    digital, 53
    distortion, 53
    rectangular, 53
    tolerance, 53
Wiggle bit, 159
Window, 19
Wired-OR, 76
Word
    binary, 26
    count, 109
    data, 26
    digital, 26
    instruction, 26
    parallel, 109
    serial, 109

XOR, 29
X state, 168, 183
XY plotter, 103

Zero
    access register, 24
    channel, 60
    delay, 60, 182